Integration with Complex Numbers

Integration with Complex Numbers

A primer on complex analysis

Aisling McCluskey

Personal Professor in Mathematics
National University of Ireland Galway

Brian McMaster

Honorary Senior Lecturer
Queen's University Belfast

OXFORD
UNIVERSITY PRESS

OXFORD
UNIVERSITY PRESS

Great Clarendon Street, Oxford, OX2 6DP,
United Kingdom

Oxford University Press is a department of the University of Oxford.
It furthers the University's objective of excellence in research, scholarship,
and education by publishing worldwide. Oxford is a registered trade mark of
Oxford University Press in the UK and in certain other countries

© Brian McMaster and Aisling McCluskey 2022

The moral rights of the authors have been asserted

Impression: 1

Published in the United States of America by Oxford University Press
198 Madison Avenue, New York, NY 10016, United States of America

British Library Cataloguing in Publication Data

Data available

Library of Congress Control Number: 2021949554

ISBN 978–0–19–284607–5 (hbk)
ISBN 978–0–19–284643–3 (pbk.)

DOI: 10.1093/oso/9780192846075.001.0001

Printed and bound by
CPI Group (UK) Ltd, Croydon, CR0 4YY

Cover image: Fedor Selivanov/Shutterstock.com

We dedicate this book to our families
with love and gratitude
and in calm acceptance of the fact that they are
never going to read more than a page or two of it.

AMcC, BMcM, October 2019

Preface

Complex analysis, more than almost any other undergraduate topic in mathematics, runs the full pure/applied gamut from the most subtle, difficult and ingenious proofs to the most direct, hands-on, engineering-based applications. This creates challenges for the instructor as much as for the very wide range of students whose various programmes require a secure grasp of some of its aspects. Its techniques are indispensable to many, but skill in the use of a mathematical tool is hazardous and fallible without a sound understanding of why and when that tool is the right one to pick up: this kind of understanding develops only by combining careful exploration of ideas, analysis of proofs and practice across a range of exercises.

Part of the challenge and the joy of teaching complex analysis is that there is no 'typical profile' for its customer base. Students from practically oriented disciplines in the physical sciences need to be able to evaluate complicated real integrals efficiently and accurately and, in many cases, complex analysis provides the most suitable techniques for enabling this; however, those techniques need to be understood in sufficient depth that their application shall be secure and dependable. Students whose primary interests lie within mathematics itself require experience in developing coherent logical arguments and then communicating them in insightful and convincing language. Between these two (relative) extremes lie various smaller sub-cohorts. Indeed, virtually everyone who is enrolled on a degree pathway within physical, engineering or mathematical sciences will encounter, during their first and second years of undergraduate study, some of the key features of complex analysis. Those whose degrees are explicitly in mathematics are likely to need many of them.

The common factors in the requirements of these diverse groups include clarity of exposition, an empathetic understanding of the difficulties that learners are likely to meet, an ample range of examples with and without specimen solutions, and a consistent determination to allow enough time for fundamental ideas to bed down into a secure foundation for confident progression. Our text offers a reader-friendly contemporary balance between idea, proof and practice, informed by several decades of classroom experience and a seasoned understanding of the backgrounds, motivation and competing time pressures of today's student cohorts. To achieve its aim of supporting and sustaining such cohorts through those aspects of complex analysis that they encounter in first- and second-year study, it also balances the competing needs to be self-contained, comprehensive, accessible and engaging—all in sufficient but not in excessive measures. In particular, it begins where most students are likely to be, and invests the time and effort that are required in order to deliver accessibility and introductory gradualness.

The principles which have guided us in constructing this text—as also in our other OUP books—are:

- Allow as much time as it takes to establish confident facility with fundamental ideas.
- Explicitly describe how solutions (that are new to the learner) are discovered, as well as how those discoveries are refined into logically secure mathematical arguments; authorise intuitive attempts, informed guesses and roughwork as steps towards the rigorously established conclusion.
- Illustrate each newly presented idea with worked examples, with exercises supported by partial solutions or hints, and with opportunities to learn by doing alongside appropriate support and encouragement.
- Be upfront in acknowledging that some arguments are difficult, and that some important topics can legitimately be omitted by substantial sections of the readership unless and until their need arises.

Our aim here is to present a primer in complex analysis—no book of modest dimensions and leisurely expository style can hope to offer any encyclopaedic overview—in which we focus toward the use of contour integration as a tool for evaluating seemingly difficult real integrals, building up from first principles the broad and thorough base of theory and skills needed to support that goal. We have felt free to include many informal sketch diagrams to give the reader a degree of visual insight into ideas and procedures and to explore intuitively how to approach a complicated problem, and thereby to encourage students to do likewise. The text is richly furnished with examples and exercises: for it remains central both to our teaching philosophy and to our classroom experience that it is by *doing mathematics* that the learner acquires secure knowledge and confidence, not simply by *reading mathematics*. There are around 120 exercises gathered up at the ends of chapters (solutions available to instructors on application to the publishers) and a broadly similar number of examples (mostly with full or sketch solutions) integrated into the text itself where they illustrate, motivate and encourage active learning. These constitute the workspace in which the apprentice analyst trains up towards craftsmanship, and where the effectiveness of a *working textbook* such as this can best be experienced.

A note to the instructor

This book is designed as a primer in complex analysis. It sets out to engage students who have no previous exposure to complex numbers and to bring them up to a good standard of confident proficiency in one major application area: that of evaluating challenging real definite integrals by associating them with complex contour integrals for which routine algorithms are available. This is a technique that we have found consistently to excite and satisfy quite diverse cohorts, and which delivers insight into the power, the applicability and something of the unexpected beauty and elegance of the subject. In addition, the careful construction and exploration of the background concepts that are needed to explain and justify this process, together with an extensive spectrum of examples and exercises aimed at developing competence and skill in their use, give the reader a broad grounding in most of the elementary aspects of complex analysis. Depending on how rapidly your classes are able to proceed, and how much background they have already acquired in cognate areas, there is—in our experience—enough material here for two first-year undergraduate semesters, or for one second-year semester.

Most of the exercises and examples that are embedded in the text come with specimen solutions complete or outlined or hinted at. Those that are gathered together at the end of most chapters constitute a suite of around 120 problems that will assist you in creating assessments for your student groups: specimen solutions for these are available to you upon application to the publishers, but not directly to your students: please see the webpage www.oup.co.uk/companion/McCluskey for how to gain access to them.

Prior knowledge that your students should ideally have before undertaking study of this material includes elementary manipulative algebra up to the binomial theorem and partial fractions, a clear intuitive understanding of the real number system, the use of simple inequalities including modulus, set theory as a language and proof techniques such as proof by induction and by contradiction, familiarity with elementary calculus (especially differentiation and integration), and basic real analysis such as convergence of sequences and series. However, we devote Chapters 1, 3 and 5 to reviewing and revising the main topics in this list so it is fully practicable to use the book for less extensively prepared cohorts. Here we also introduce and develop, so far as is needed, two topics that are often missing from students' backgrounds—partial differentiation and improper integrals.

A note to the student reader

Whatever your reasons may be for wanting or needing to study integration involving complex numbers, it is likely that your intended goals include the following:

- An understanding of how so-called 'imaginary' numbers can help you in the study of real-world mathematical problems;
- The ability to handle a range of apparently difficult real integration questions by converting them into relatively straightforward complex integrals;
- Competence and confidence in handling complex integrations and in explaining your arguments and conclusions to other people; and
- A solid understanding of the basic ideas and algorithms that underlie these matters.

This book is designed to help you achieve this and more. In particular, it allows time to explore and develop the initial ideas that need to be securely grasped before embarking on more challenging issues, it incorporates informal discursive sections to build up your intuitive feel for concepts that will guide and inform you in working with their proper mathematical formulations, and it steps aside from time to time into roughwork paragraphs that provide a kind of apprenticeship experience through which you can look 'over the shoulder' of a more experienced practitioner and 'under the bonnet' of a complicated problem.

It will help you to get quickly into the interesting parts of this subject if you have already acquired basic knowledge and technique in 'real' calculus and analysis, especially concerning the ideas of sets, sequences and series and of the differentiation and integration of real functions. However, we revise these topics in Chapters 1, 3 and 5, together with accounts of two matters that are sometimes missing from the background of intending readers—partial differentiation and improper integrals. Chapter 2 offers an introduction to complex numbers from the very beginning, and Chapter 4 initiates the development of calculus in the complex setting. If you happen to be thoroughly familiar with these matters already then you can start the true business of the book from Chapter 6.

Our emphatic advice to you is to try every exercise or example as you meet it, and then either to check your answer with a model solution if the book or your lecture course makes one available or to see if your account convinces a tutor or a fellow student. There are over 120 exercises compiled at the ends of our chapters, and a comparable number of examples—mostly with full solutions—embedded in the text itself. Try them all! Mathematics is not a subject that you can learn by passively reading: it is a practical, performance-based discipline whose skills can only be acquired and sharpened by doing. The more mathematics you do—even do wrongly, provided that you find out the weak points in your work and strengthen them for the next occasion—the deeper your understanding and your skill base will become, and the greater the enjoyment you will derive from doing more.

Contents

Background part A
set, sequence, series

1.1 Introduction

Anyone who wishes to work with aspects of calculus or analysis—limits, continuity, series, differentiation, integration—in the context of *complex* numbers will be well advised to have first studied the corresponding ideas in the context of *real* numbers. This is because most of us feel the real number system to be much closer to everyday experience than its complex cousin, therefore leaving us freer to focus on acquiring skill in and familiarity with these rather challenging notions without needing to worry about the nature, or even the existence, of the numbers underlying what we are doing. All the same, it is important to realise that only quite a limited grasp of real calculus/analysis is necessary before beginning to diversify into studying its complex clone: for the basic ideas that we need in the two areas are virtually identical, and the same can be said about many of the results that are built upon them.

What we seek to do in Chapters 1, 3 and 5 of the present text is to summarise as succinctly as possible the pieces of real calculus and real analysis that we should like the reader to have met *before* starting to work on the main material. These background chapters do not set out to be anything like complete accounts of the topics that headline them, and their intended purpose varies from one learner to another. Depending on the experience, the time constraints and the motivation of individual readers, they are designed

1. *either* briefly to remind the reader who has already achieved a good grasp of basic real analysis about the particular concepts, results and notations that we shall directly use in later chapters,

2. *or* to encourage the less-well-versed reader to try to make the time to acquire an understanding and a modest degree of skill in the concepts, results and notations listed (possibly by referring to one of the texts named under 'Suggestions for Further Reading'),

3. *or, at absolute minimum,* to list—as clearly as time and space permit us—the ideas of real analysis upon which this account of the development of complex analysis is going to be built.

Integration with Complex Numbers. McCluskey and McMaster, Oxford University Press.
© Brian McMaster and Aisling McCluskey (2022). DOI: 10.1093/oso/9780192846075.003.0001

1.2 **Revision 1: sets**

We shall begin with a very quick review of the language and notation of set theory insofar as they abbreviate and streamline what we need to say later concerning, for instance, collections of complex numbers.

1.2.1 **Definitions**   A *set* is any well-defined collection of distinct objects. 'Well-defined' in this context means that for any set A and any relevant object a it must be unambiguously decidable whether a *is* or *is not* one of the objects that belong to A. 'Distinct' means that if a is an object that belongs to a set A and b is an object that belongs to A, then it must be unambiguously decidable whether $a = b$ or $a \neq b$. The notation

$$a \in A$$

says that A is a set and that a is one of the objects that belong to it (that is, that a is an *element* of the set A). The notation

$$a \notin A$$

says that a is *not* one of the elements of the set A. The special symbols $\mathbb{N}, \mathbb{Z}, \mathbb{Q}, \mathbb{R}$ and $\mathbb{C}$ always denote the set of *positive integers*, the set of all *integers* (whole numbers positive, negative or zero), the set of *rationals* (the numbers that can be expressed exactly as one integer divided by another), the familiar set of *real* numbers, and the set of *complex numbers* (which, naturally, we shall discuss in some detail later on). So, for instance, the shorthand

$$x \in \mathbb{C}, x \notin \mathbb{R}$$

is simply a quick way to write that x is a complex number but that it is *not* a real number.

Finite sets can be simply and conveniently represented by enclosing a list of their elements within curly brackets; for instance, $\{2, 5, 7, 11\}$ is the set of prime divisors of 3080. Similar notations can *with caution* be employed for some infinite sets; no confusion is likely to arise if we write $\{1, 2, 3, 4, \ldots\}$ for $\mathbb{N}$ or $\{\ldots - 2, -1, 0, 1, 2, 3, \ldots\}$ for $\mathbb{Z}$. On the other hand, *more seriously infinite* sets such as $\mathbb{R}$ cannot be written in this way.

1.2.2 **Definitions**   If A and B are sets such that *every* element of A is also an element of B then we call A a *subset* of B and write $A \subseteq B$ (or, equivalently, $B \supseteq A$). We note the extreme cases: $A \subseteq A$ is always true, and so is $\emptyset \subseteq A$ where $\emptyset$ denotes the 'empty set', that is, the set that has no elements at all. If $P(x)$ is any meaningful statement about the typical element x of a set A, then the symbols

$$\{x \in A : P(x)\} \quad or \quad \{x \in A \mid P(x)\} \quad or \quad \{x : x \in A, P(x)\} \quad or \quad \{x \mid x \in A, P(x)\}$$

all denote the subset of A that comprises exactly those elements x of A for which $P(x)$ is a true statement. For instance, $\{x \in \mathbb{R} : x \notin \mathbb{Z}\}$ is the set of all real numbers that are not integers, $\{x \in \mathbb{R} : -1 < x < 2\}$ is the 'open interval' of all real numbers that lie strictly between -1 and 2, often written more briefly as $(-1, 2)$, and $\{z \in \mathbb{C} \mid z^8 = 1\}$ consists of all the complex eighth roots of unity (another matter to be discussed later). Of course, $\{x \in A : x \neq x\}$ is just $\emptyset$.

1.2.3 **Definitions**   For any sets A and B we define their *union* $A \cup B$, their *intersection* $A \cap B$ and their *set difference* $A \setminus B$ as follows:

$$A \cup B = \{x : x \in A \text{ or } x \in B \text{ (or both)}\},$$

$$A \cap B = \{x : x \in A \text{ and } x \in B\},$$

$$A \setminus B = \{x : x \in A \text{ but } x \notin B\}.$$

In the case where $B \subseteq A$, the set difference $A \setminus B$ is usually called the *relative complement of B in A* or, more briefly, the *complement of B* if the context makes it obvious which background set A is being considered.

The ideas of union and intersection extend routinely to deal with more than two sets—indeed, we can define the union of any family of sets (even an infinite family) to comprise all the objects that are elements of *at least one* set in the family, and the intersection of any family of sets to comprise all objects that belong to *every* set in the family.

There are a large number of 'rules' that facilitate combining sets via unions, intersections and complements in an almost mechanical way, but we are unlikely to need them very much (and will draw attention on an *ad hoc* basis to any that do turn out to be useful to us).

1.2.4 **Note**   A non-empty subset of $\mathbb{R}$ that has no breaks in it is called an *interval*. Intervals that 'stretch off to infinity' or 'to minus infinity' are called *unbounded*, and those that do not are called *bounded*. The point or points (if there are any) that lie on the borderline of an interval are called its *endpoints*, and they themselves may or may not belong to the interval as elements. These interval subsets of $\mathbb{R}$ turn up so often that we should make the above rather informal descriptions fully precise, and also take note of the standard symbols for intervals of the various types (here, a and b are real numbers and denote the endpoints):

1. $(a, b) = \{x \in \mathbb{R} : a < x < b\}$
2. $[a, b] = \{x \in \mathbb{R} : a \leq x \leq b\}$
3. $[a, b) = \{x \in \mathbb{R} : a \leq x < b\}$
4. $(a, b] = \{x \in \mathbb{R} : a < x \leq b\}$
5. $(a, \infty) = \{x \in \mathbb{R} : a < x\}$
6. $[a, \infty) = \{x \in \mathbb{R} : a \leq x\}$
7. $(-\infty, b) = \{x \in \mathbb{R} : x < b\}$

8. $(-\infty, b] = \{x \in \mathbb{R} : x \le b\}$
9. $(-\infty, \infty) = \mathbb{R}$
10. $[a, a] = \{x \in \mathbb{R} : a \le x \le a\} = \{a\}$

Intervals such as 1, 2, 3 and 4 (and, technically, 10) are the bounded intervals, whereas the others are unbounded. Numbers 1, 5, 7 and 9 are *open intervals* and numbers 2, 6, 8, 9 (and, technically, 10) are *closed intervals*. Note that item 10 is often considered not to be an interval at all, whereas some texts classify it as a *degenerate interval*.

1.3 Revision 2: sequences

1.3.1 Definitions   A *sequence* is an unending succession of numbers in a particular order (a first, a second, a third…). We call it an *integer sequence* if all its terms are integers, a *real sequence* if all its terms are real numbers, and so on. Standard ways of writing down a sequence include

$$(a_1, a_2, a_3, a_4, \ldots)$$

and

$$(a_n)_{n \in \mathbb{N}}$$

and even just

$$(a_n).$$

In order fully to describe a particular sequence, we need to include instructions on how to determine the typical term[1] a_n, and that usually entails *either* an explicit formula for a_n *or* a recursive procedure for creating it from earlier terms in the sequence. A classic example of the latter is the Fibonacci sequence $(1, 1, 2, 3, 5, 8, 13, 21, \ldots)$ in which the recursive pattern is that $a_{n+2} = a_{n+1} + a_n$ (that is, each term from the third one onwards is defined by adding the two that immediately precede it); an explicit formula for the nth Fibonacci number as a function of n *can* be found, but it is far from obvious at first sight.

Some sequences, but by no means all of them, settle towards an 'equilibrium state' or 'limiting value' as we scan further and further along the list of their terms.

---

[1] Unless, of course, the pattern of the first few terms is *so* obvious that nothing further is required: for instance, in the case of the sequence

$$\left(1, \frac{1}{2}, \frac{1}{4}, \frac{1}{8}, \frac{1}{16}, \ldots\right)$$

it is surely unnecessary to say explicitly that the nth term is $\dfrac{1}{2^{n-1}}$.

When this does occur, it is the most important thing to know about that particular sequence. Here follows the precise definition: recall that the modulus $|x|$ of a real number x is the physical size of x, disregarding whether it is positive or negative, that is,

$$|x| = \begin{cases} x & \text{if } x \geq 0, \\ -x & \text{if } x < 0. \end{cases}$$

1.3.2 **Definitions**   A sequence $(a_n)_{n \in \mathbb{N}}$ is said to *converge* or to *be convergent* to a *limit L* (as $n \to \infty$) if, for each $\varepsilon > 0$, there is a positive integer n_ε such that

$$n \geq n_\varepsilon \Rightarrow |a_n - L| < \varepsilon.$$

Instead of $a_n \to L$ as $n \to \infty$, the alternative notation $L = \lim_{n \to \infty} a_n$ is often used. A sequence that is not convergent (that is, does not possess a limit) is said to *diverge* or to *be divergent*.

This is not an easy definition to assimilate; we shall use it as little as possible in the text, but we strongly recommend that you make an effort (again we flag up the additional reading resources listed at the end of the book) to build up some degree of confidence in what it means and in how to use it. It may be helpful to think of ε as the *tolerance of error* that can be permitted in some envisaged application, so that the definition of convergence to L effectively says:

> no matter how small the tolerance ε has been set, we can make a_n a good approximation to L, that is, one whose error or inaccuracy $|a_n - L|$ is smaller than that allowed tolerance ε, just by going far enough along the sequence, that is, by making sure that the label n on the term a_n is *at least* some *threshold value* n_ε (which, of course, probably depends on how small the tolerance ε is required to be).

Despite the somewhat intimidating appearance of the definition of sequence limit, many of the results that follow from it look extremely natural (and, indeed, have quite straightforward proofs). For example, if a sequence (c_n) has been created by adding together two convergent sequences (a_n) and (b_n) (that is, $c_n = a_n + b_n$ for every n), then the fact that (c_n) also converges, and that its limit is just the sum of the two limits of its summands:

$$\lim(a_n + b_n) = \lim a_n + \lim b_n \quad (\text{as } n \to \infty \text{ in each case})$$

looks like nothing more than common sense. Likewise, the limits of differences, products, quotients and so on of convergent sequences turn out to be exactly what common sense expects them to be (provided always that no attempt is made to divide by zero, of course).

1.3.3 **Definition**   A *subsequence* of a given sequence

$$(a_n) = (a_1, a_2, a_3, a_4, \ldots)$$

is any sequence that is formed by selecting an unending list of some of its elements in their original order

$$(a_{n_1}, a_{n_2}, a_{n_3}, a_{n_4}, \ldots)$$

(where $n_1 < n_2 < n_3 < n_4 < \ldots$ to ensure that the order of the selected terms is as it was in the parent sequence). The (admittedly rather cumbersome) standard notation for such a subsequence is

$$(a_{n_k})_{k \in \mathbb{N}}.$$

We need to flag up three less obvious but important results.

1.3.4 **Theorem**

- Every convergent sequence is bounded (that is, we can find some positive constant K so that every term of the sequence lies between K and $-K$).
- If a sequence converges to a certain limit, then every one of its subsequences also converges (and to that same limit).
- Every *bounded* real sequence possesses a convergent subsequence. (Note that this result is known as the **Bolzano-Weierstrass** theorem.)

(Incidentally, the first and second parts of this *portmanteau* theorem give us the two standard ways of trying to show that a given sequence is *divergent*: we should either try to prove that it is not bounded, or else seek two subsequences of it that possess *different* limits. If we succeed in either of those enterprises, the sequence cannot have been convergent.)

1.4 **Revision 3: series**

Informally, a series is the problem we run into once we aspire to add all the terms of an (infinite) sequence. Since addition simply doesn't work like that (for it only makes sense to add a *finite* list of numbers) this is not good enough to be a proper definition, but it does virtually tell us how to tighten the logic in a way that does make good sense but still describes the original intention. To try to add 'all' the terms in an unending sequence, we should instead add the first n of them, get (or estimate) that nth *partial* sum, and look for a limit of this quantity as n becomes larger and larger.

1.4.1 Definition   For any given real sequence $(a_k)_{k \in \mathbb{N}}$ let $(s_n)_{n \in \mathbb{N}}$ be the other sequence defined by $s_1 = a_1, s_2 = a_1 + a_2, s_3 = a_1 + a_2 + a_3$ and, in general,

$$s_n = \sum_{k=1}^{n} a_k = a_1 + a_2 + a_3 + \ldots + a_n.$$

The *series* $\sum_1^{\infty} a_k$ is the pair of sequences (a_k) and (s_n). It is said to be a *convergent* series if (s_n) is convergent (and then the limit of (s_n) is called the *sum* or, more formally, the *sum to infinity* of the series); if (s_n) has no limit, the series is said to be *divergent*. The two constituent sequences (a_k) and (s_n) are called the *sequence of terms* and the *sequence of partial sums* of the series.

Common experience is that

1. it can be very difficult to determine the sum of a series, but

2. usually that does not matter very much, since (most of the time) it is more important to know whether a series does or does not converge than to evaluate any actual sum,

3. and in consequence we will benefit by gathering together a number of 'convergence tests' that focus simply on determining whether a series does converge or not.

4. In this regard, series whose terms are all non-negative are easier to work with: because such a series converges *if and only if* its sequence of partial sums is bounded above (that is, if and only if we can find some constant that is bigger than every partial sum).

1.4.2 Comparison tests

Provided that all the terms of two series $\sum_1^{\infty} a_k$ and $\sum_1^{\infty} b_k$ are non-negative:

1. If $a_k \le b_k$ for every k, and the 'bigger' series $\sum_1^{\infty} b_k$ converges, then also the 'smaller' series must converge.

2. To say the same thing in slightly different language, if $a_k \le b_k$ for every k and the 'smaller' series $\sum_1^{\infty} a_k$ diverges, then also the 'bigger' series must diverge.

3. If we can find a particular positive integer n_0 such that $a_k \le b_k$ for every $k \ge n_0$ then, again, if the 'bigger' series $\sum_1^{\infty} b_k$ converges, then also the 'smaller' series must converge, and if the 'smaller' series $\sum_1^{\infty} a_k$ diverges, then also the 'bigger' series must diverge.

4. If we can find a particular positive real number C such that $a_k \le C b_k$ for every k then, once again, if the 'bigger' series $\sum_1^{\infty} b_k$ converges, then also the 'smaller' series must converge, and if the 'smaller' series $\sum_1^{\infty} a_k$ diverges, then also the 'bigger' series must diverge.

5. The **limit comparison test**: If the ratio $\dfrac{a_k}{b_k}$ converges to a *non-zero* limit (and none of the terms a_k, b_k is zero) then the two series $\sum_1^\infty a_k$ and $\sum_1^\infty b_k$ will either *both* converge or *both* diverge.

(The proofs of all these tests use point 4 of 1.4.1 and very little else.)

Two of the most useful series tests (again, for series whose terms are all non-negative) concern the limiting value of either the growth rate (as described in 1.4.3) or the nth root of the nth term:

1.4.3 Theorem

1. **The ratio test (of d'Alembert)** Suppose that $\sum a_n$ is a series of positive terms, and that the growth rate

$$\frac{a_{n+1}}{a_n}$$

converges to a limit L (as $n \to \infty$). Then
   (a) $L < 1$ implies that the series is convergent,
   (b) $L > 1$ implies that the series is divergent.

2. **The nth root test** Suppose that $\sum a_n$ is a series of non-negative terms, and that

$$\sqrt[n]{a_n}$$

converges to a limit L (as $n \to \infty$). Then
   (a) $L < 1$ implies that the series is convergent,
   (b) $L > 1$ implies that the series is divergent.

Proofs of these can be found in almost any textbook on the elements of real analysis. Note that if the limit in question is *exactly* 1, then neither test tells us anything useful.

Turning to the question of how to assess convergence for a series whose terms are a mixture of positives and negatives, we should revise one rather specialised device (1.4.4) and a much more universal approach (1.4.5 and 1.4.7). See any standard analysis textbook for proofs if desired.

1.4.4 **The alternating series test**

Suppose that (a_n) is a decreasing[2] sequence of positive numbers that converges to 0. Then the series

$$a_1 - a_2 + a_3 - a_4 + a_5 - a_6 + \ldots = \sum_1^\infty (-1)^{n+1} a_n$$

converges.

---

[2] That is, $a_n \geq a_{n+1}$ for every n.

1.4.5 Definitions   A series $\sum_1^\infty a_n$ is called *absolutely convergent* if the 'modulussed' series $\sum_1^\infty |a_n|$ is convergent. A series that is convergent but *not* absolutely convergent is said to be *conditionally convergent*.

1.4.6 Examples

- The *alternating harmonic series* $\sum_1^\infty \dfrac{(-1)^{n+1}}{n} = 1 - \frac{1}{2} + \frac{1}{3} - \frac{1}{4} + \frac{1}{5} - \frac{1}{6} + \ldots$ is conditionally convergent.

- For any constant $t \in (-1, 1)$, the *geometric series* $\sum_1^\infty t^{n-1} = 1 + t + t^2 + t^3 + t^4 + \ldots$ is absolutely convergent, and its sum is $\dfrac{1}{1-t}$.

- For any constant $u > 1$, the series

$$\sum_1^\infty n^{-u} = 1 + \left(\frac{1}{2}\right)^u + \left(\frac{1}{3}\right)^u + \left(\frac{1}{4}\right)^u + \left(\frac{1}{5}\right)^u + \ldots$$

is (absolutely) convergent.

1.4.7 Theorem   Every absolutely convergent series is convergent.

One of the startling things about a *conditionally* convergent series is that we can, simply by changing the order in which we add up its terms, make it converge to *any sum-to-infinity whatsoever* that we choose.

One of the reassuring things about an *absolutely* convergent series is that it cannot behave badly like that: no matter in what order you gather its terms, the partial sums will still converge to exactly the same limit (again, see any standard analysis textbook if you want to see how to prove these assertions):

1.4.8 Theorem   Suppose that $\sum_1^\infty a_n$ is absolutely convergent and that S denotes its sum. Let $\sum_1^\infty b_n$ be any rearrangement of $\sum_1^\infty a_n$ (that is, the sequence of terms $(b_n)_{n \in \mathbb{N}}$ consists of the same numbers as does $(a_n)_{n \in \mathbb{N}}$ but in a different order). Then $\sum_1^\infty b_n$ is also absolutely convergent, and its sum is S also.

The most useful class of series, and the one you are most likely to have worked with before, is what is called the class of power series. A *power series* is a series of the form[3]

$$a_0 + a_1 x + a_2 x^2 + a_3 x^3 + \ldots + a_n x^n + \ldots = \sum_0^\infty a_n x^n$$

---

[3] The slightly more general form of power series

$$a_0 + a_1(x - a) + a_2(x - a)^2 + a_3(x - a)^3 + \ldots + a_n(x - a)^n + \ldots = \sum_0^\infty a_n(x - a)^n$$

that we often meet can be turned into this shape by the simple substitution $X = (x - a)$.

where x is a variable and the constants a_n are called its *coefficients*. So, informally speaking, a power series 'looks like' a polynomial, except that it has 'infinite degree'.[4]

(Notice that we start off a power series $\sum_0^\infty a_n x^n$ at $n = 0$ instead of at $n = 1$, to allow it to have a constant term $a_0 = a_0 x^0$.)

Whether or not a power series converges and, if it does, what its sum is, naturally depend on the value of x. For almost every power series $\sum_0^\infty a_n x^n$ we can find a number called its *radius of convergence* (let's denote it by r) such that

1. $\sum_0^\infty a_n x^n$ is absolutely convergent if $-r < x < r$,
2. $\sum_0^\infty a_n x^n$ is divergent if $x < -r$ and also if $r < x$,
3. (and it is usually harder to decide whether it does or does not converge when $x = \pm r$, but also usually not so important).

The only exceptions to this pattern are (i) there are some power series $\sum_0^\infty a_n x^n$ that *only* converge when $x = 0$, and (ii) there are quite a few that converge absolutely for *every* real value of x. The radius of convergence is defined to be 0 for the first type of special case, and is said to be infinite for the second.

Three instances of the 'second type of special case' are especially useful for our purposes:

1.4.9 **Theorem** For every (real) number x:

1. the series

$$1 + x + \frac{x^2}{2!} + \frac{x^3}{3!} + \frac{x^4}{4!} + \dots$$

is absolutely convergent to e^x,

2. the series

$$x - \frac{x^3}{3!} + \frac{x^5}{5!} - \frac{x^7}{7!} + \dots$$

is absolutely convergent to $\sin x$, and

3. the series

$$1 - \frac{x^2}{2!} + \frac{x^4}{4!} - \frac{x^6}{6!} + \dots$$

is absolutely convergent to $\cos x$.

---

[4] Although that sentence is informal, it is a pretty good pointer to how power series actually behave in practice; for instance, it turns out that you can usually differentiate a power series one term at a time, just as if it really were a polynomial! We shall return to this later, when we revise functions and differentiation.

2 What are complex numbers?

Computant ergo sunt?

Hopefully you—the reader—will forgive us if we begin this chapter by disrespecting the question embedded in its title. For in fact, it hardly matters at all what complex numbers are; what matters is how they behave, how we can work with them, and what they can do for us.

To get some insight into the point that we are seeking to make here, try to imagine a discussion between, say, Isaac Newton, Gottfried Leibniz and Auguste Cauchy on 'what the positive integers are'. Newton says they are the numbers one, two, three, four and so on. Leibniz disagrees—they are eins, zwei, drei, vier und so weiter. For Cauchy, they are une, deux, trois, quatre, cinq... And already you should be rejecting this scenario as patently ludicrous: these folk certainly know far too much about their discipline to believe for an instant that they are disagreeing about the numbers themselves: their divergence of opinion is only about what names to give these integers for purposes of communication—it is not about the numbers as numbers.

Very much the same issue arises if you imagine an exchange of correspondence between, let us say, a first-century Roman governor, a mediaeval Arabic scholar and a modern IT specialist. The Roman writes that the positive integers are I, II, III, IV, V and so on; for the Arabic scholar (as for us) they are 1, 2, 3, 4, 5... while the IT guy possibly chooses a binary notation beginning with 1, 10, 11, 100, 101, 110... Are they disagreeing as to what the positive integers are? No—the nature of their conflict is merely about what symbols it is convenient to use in order to write them down and work with them. This is certainly an important issue in its own right, as you are likely to find out if you try to do multiplication or long division using Roman numerals! However, any question about the numbers themselves—how many prime numbers do you encounter if you count off your fingers and toes? Is it always true that when you want to multiply two numbers together, then it makes no difference which one you start with? Can there be two positive integers in which the square of the bigger one is exactly twice the square of the smaller one?—will receive the same answer from all these debaters if they have sufficient time and curiosity to think hard enough about it.

The reason why we think it necessary to make a fuss about such an obvious detail concerns the question of existence. Our shared classroom experience is that

Integration with Complex Numbers. McCluskey and McMaster, Oxford University Press.
© Brian McMaster and Aisling McCluskey (2022). DOI: 10.1093/oso/9780192846075.003.0002

when you square a number (that is, multiply it by itself) then the answer is zero if the number itself is zero, is positive if your starting number is positive, and is also positive if your starting number is negative. In other words, the square of any of the real numbers (effectively, those that can be classified as positive or negative or zero) cannot be negative. Unfortunately, this conclusion becomes abbreviated with use: 'there exists no number whose square is negative' or 'square roots of negatives don't exist'. Of course, these are perfectly valid abbreviations if the only numbers you need to work with are real numbers—but our ambition here is broader than that.

Questions of existence have created trouble several times in the history of mathematics. Classical Greece, despite the extraordinary sophistication of its geometry, had great difficulty accepting the legitimacy of 'numbers' such as $\sqrt{2}$ that could not be expressed exactly as the ratio of two whole numbers. Mediaeval Europe rejected for centuries the idea that negatives or zero should be accepted as 'numbers' in the ordinary sense of the term. Yet with the passing of time and the accumulation of experience, two things came to be understood: (i) these rogue numbers obeyed very much the same operating rules as did the fully domesticated counting and measuring numbers that we had been using for millennia, and (ii) accepting the rogues into the family of socially acceptable numbers made certain calculations and arguments easier and, in the longer run, led to new and exciting insights and techniques.

A running thread through this historical development of insight as to 'which types of number exist' concerned the growing usefulness of basic algebra in formulating (and hopefully in subsequently solving) real-world problems. If the only numbers whose existence you accept are positive integers, then even such a simple algebraic equation as $2x = 5$ has no solution; but once it becomes acknowledged that positive fractions are perfectly legitimate objects, then all equations of the general form $ax = b$ (with a and b being positive integers) become solvable at a stroke. Equations such as $x + x = x$ and $x + 10 = 7$ have no numerical solution unless and until you accept that negatives and zero qualify as numbers; but once that nettle is grasped, it becomes a matter of simple routine to solve any equation of the form $ax + b = c$ (provided only that $a \neq 0$). For so long as we are reluctant to tolerate so-called irrationals such as $\sqrt{2}$ and $3 + \sqrt{5}$ as numbers, then equations as pedestrian as $x^2 = 2$ and $x^2 - 6x + 4 = 0$ cannot be solved; yet once square roots of all positives are admitted to the number family, a large proportion of these quadratic equations become solvable in a straightforward and uniform manner.

And this is the point at which the first (and probably the weakest) reason to 'choose to believe in the existence of square roots of negatives' shows up. The quadratic formula—the algebra that rearranges the question

$$\text{if } ax^2 + bx + c = 0, \quad \text{what can } x \text{ be?}$$

into the answer

$$x = \frac{-b \pm \sqrt{b^2 - 4ac}}{2a} \ !$$

does not appear to care whether the square root on the top line 'exists'. Yet so long as we only admit real numbers as being legitimate objects for arithmetic, it seems to present us with two distinct cases: if $b^2 - 4ac$ is positive, it yields two solutions for x; however, if $b^2 - 4ac$ is negative, no solution 'exists'. Thus, we reach—and not for the first time in history—a point at which security and efficiency seem to want to follow different paths: rejecting square roots of negatives from consideration keeps our solutions safely within the field of the real numbers which we think we have thoroughly understood for centuries, but at the cost of being able to solve only about half (broadly speaking) of quadratic equations; on the other hand, accepting them into the realm of calculations would give us the power to solve all quadratics—and solve them all by exactly the same procedure—but paying the discomfort price of working with objects that are seriously further away from our everyday experience. Of course, this price would only be worth paying if *firstly* we can develop a decent notation for writing down and working with these new, semi-mythical 'numbers', *secondly* we manage to assure ourselves that they obey (at least most of) the standard familiar rules of arithmetic and *thirdly* they lead us on to fresh insights and applications—better ones than merely boasting that we can now solve all quadratics!

Naturally, all three of these conditions were long ago found to be satisfied: otherwise, texts such as this would never be written. Before starting to gather evidence for this assertion, though, we ought to flag up two smaller points that arise from the above discussion. Firstly, in connection with the imagined disagreement between the IT developer and the Arabic scholar as to whether decimal presentation or binary (or hexadecimal) notation is the better symbol system for representing positive integers, the pragmatic answer is that neither is globally better—it depends on what you plan to do with them afterwards: most of us take the decimal option if we intend pen-and-paper calculation or use of a normal calculator, whereas digital computers and those who work most directly with their internal structures tend to use binary or binary-derived number representation. In much the same way, we shall see that there is no single 'best' way to denote 'complex numbers'—numbers of the kind that arise once we 'choose to believe' in square roots of negatives. Rather, there are several, each with its own local benefits, and it will pay us dividends to be able to switch from one to another within an argument in order to combine insight or imagery from more than one notational system.

Secondly—and this is a point that is not really about mathematics at all, but about the world view of those who use it—let us take a brief look at the question of the existence of complex numbers. Mathematics is pretty good at establishing beyond any reasoned doubt that certain things do not exist—fractions (in the ordinary sense) whose squares equal 2 or 3 or 5 exactly, a largest prime number, a ruler-and-straight-edge construction for trisecting any given angle, a differentiable real function that is not continuous—these and many other conjectural entities can be shown non-existent, often by demonstrating that the very idea that they might exist contradicts either the idea itself or some established, universally accepted notion. On the other hand, proofs of existence tend to be constructions based on accepting *a priori* the existence of other objects that are arguably simpler or less controversial, which in turn depends on accessing still simpler objects, and which in turn... And although this pyramidal character of mathematical constructabili-ty is one of its greatest strengths, it can sometimes seem to distance us from the emotional, the psychological certainty of the reality of the things that are being constructed. There is no help for this: none of us has ever seen the number 8, the number π, the number -1500, although we have repeatedly seen symbols that represent them or letters that make up a name of such a symbol. Why then does nobody lose sleep over such anxieties? The answer appears to be that we have col-lectively learned to trust them; we have seen them used reliably, time after time, in deciding how many plates to set out for our dinner guests, how much paint to buy before recoating the cylindrical oil tank, how long it will take to balance our overdrawn bank account. It is the sheer utility, the consistent reliability of our use of such familiar numbers that reassures us psychologically that they are 'there' in some deep, almost platonic sense, more than does any formal logico-mathematical construction plan for their creation.

And exactly the same thing is true of the complex numbers. They play their part in sensible, predictable, useful calculations: therefore (we believe that) they are. *Computant ergo sunt!*

2.1 How do we handle them?

2.1.1 An 'algebraic' definition and notation   Complex numbers are symbols of the form $a + bi$, where a and b are real numbers and i is just a symbol.[1] They are to be added and multiplied exactly as everyone's previous experience of elementary algebra says *with the sole exception* that every time i^2 appears in a calculation, it is to be replaced by -1. (No distinction is made between $a + bi$ and $a + ib$: they are regarded as being the same symbol, differently typed.)

2.1.2 Discussion   So, for instance, $3 - 0.5i$ and $-7 + 6i$ are complex numbers, and $3 - 0.5i + (-7 + 6i) = -4 + 5.5i$; furthermore,

$$(3 - 0.5i)(-7 + 6i) = -21 + 3.5i + 18i - 3i^2$$

$$= -21 + 3.5i + 18i - 3(-1) = -21 + 3.5i + 18i + 3 = -18 + 21.5i.$$

Again, if $w = 4 - 18i$ and $z = 2 + 9i$ then $w + z$ calculates immediately to $6 - 9i$ while wz calculates easily to $8 - 162i^2 + 36i - 36i = 8 + 162 + 0i = 170 + 0i$.

Two important details of notation come to light as we finalise the last exam-ple. Firstly, in accordance with common experience, $0i$ shall be interpreted as

---

[1] Engineering texts generally use the letter j here instead of the letter i (which they often need to reserve for other purposes, e.g. electrical current).

0, and $170 + 0$ as 170. More generally, *any* real number t can be interpreted as (that is, is equal to) $t + 0$ (of course) and therefore as $t + 0i$, that is, each real number is a special case of a complex number. Going a little further, the way in which these 'special' complex numbers add and multiply

$$(t + 0i) + (u + 0i) = t + u + (0 + 0)i = t + u + 0i,$$

$$(t + 0i)(u + 0i) = tu + 0ti + 0ui + 0i^2 = tu + 0(t + u)i - 0 = tu + 0i$$

precisely matches how the corresponding real numbers that they 'shadow' add and multiply. So the complex number system incorporates not only the real numbers themselves, but also the way in which real addition and real multiplication work.

Secondly, it is standard practice to use a single letter to stand in for any complex number instead of writing it out in full. This is useful in particular when we want to highlight general behaviour patterns—for instance, in order to draw attention to the fact that the sum and the product of two complex numbers does not depend on which of the two you write down first, it is quicker and easier to write

$$\begin{cases} w + z = z + w \\ wz = zw \end{cases} \quad \text{for all complex numbers } z, w$$

rather than the equally correct but cumbersome

$$\begin{cases} a + bi + (c + di) = c + di + (a + bi) \\ (a + bi)(c + di) = (c + di)(a + bi) \end{cases} \quad \text{for all complex numbers } a + bi, \ c + di.$$

(Indeed, it is also quicker and easier to read.)

It is now entirely routine (but quite tedious) to verify that all the usual arithmetical rules—the so-called *field axioms*[2]—that are obeyed in real addition and real multiplication equally govern addition and multiplication of complex numbers as we have just described them. By way of illustration we shall check just one—the one referred to as the *associative law of multiplication*:

2.1.3 Proposition   For any complex numbers u, v and w we have $u(vw) = (uv)w$.

Proof   We can begin by writing u, v and w out in full, say: $u = a+bi$, $v = c+di$, $w = e + fi$, where a, b, c, d, e and f are real numbers. Then $vw = (ce - df) + (cf + de)i$, and so

---

[2] See Chapter 11 for a full list of these.

$$u(vw) = (a+bi)((ce-df)+(cf+de)i) = (ace-adf-bcf-bde)+(acf+ade+bce-bdf)i.$$

Next, $uv = (ac - bd) + (ad + bc)i$, so

$$(uv)w = ((ac-bd)+(ad+bc)i)(e+fi) = (ace-bde-adf-bcf)+(acf-bdf+ade+bce)i.$$

At this point, all we need to do is to observe that the last two display lines result in exactly the same answer.   ■

Recalling our promise that alternative descriptions would bring additional insights, let us now set out a different (but compatible) approach:

2.1.4 A 'geometric' definition and notation   Complex numbers are simply points in the coordinate plane, and the most natural way to represent such an object is to state its x- and y-coordinates: (a, b) (where a and b are reals) is a typical complex number. They are to be added in the natural component-wise fashion

$$(a, b) + (c, d) = (a + c, b + d)$$

but are to be multiplied according to a procedure that, at first sight, looks rather bizarre:

$$(a, b)(c, d) = (ac - bd, ad + bc).$$

2.1.5 Discussion   In this notation, we check the sum and product of a couple of complex numbers (= points in the plane): $(3, -0.5)+(-7, 6) = (3 + (-7), -0.5 + 6)$ $= (-4, 5.5)$

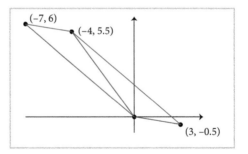

and

$$(3, -0.5)(-7, 6) = (3(-7) - (-0.5)6, 3(6) + (-0.5)(-7)) = (-21 + 3, 18 + 3.5)$$
$$= (-18, 21.5).$$

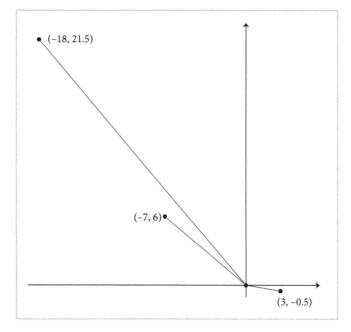

(We have added a few coloured lines to the sketch diagrams for later use. You might already want to compare these calculations with those at the start of paragraph 2.1.2 and, if you do so, you will soon realise that although we are writing different symbols, we are not really doing different things! More on this later also.)

Notice the behaviour of the 'special' complex numbers that are points on the horizontal axis, in other words, those that have zero as their second coordinate:

$$(a, 0) + (c, 0) = (a + c, 0 + 0) = (a + c, 0),$$

$$(a, 0)(c, 0) = (a(c) - 0(0), a(0) + 0(c)) = (ac, 0).$$

We see that they add and multiply as if the second coordinate were absent. So, at least as regards their basic arithmetic of adding and multiplying, the points on the x-axis behave exactly like real numbers, and (again) the complex numbers as we are now describing them contain a structure that looks—at least, up to this point—identical to the real number system.

We do not have common sense as a maxim to fall back on this time, so the verification that all the field axioms are satisfied by the second proposed definition is more evidently necessary than it was for the first, but it is just as routine (and just as tedious). Here as illustration is a verification of the *distributive law*—the rule in arithmetic/algebra that permits us to multiply out brackets. (We are again opting to let single letters stand in for arbitrary complex numbers.)

2.1.6 Proposition   For any complex numbers u, v and w we have $u(v + w) = uv + uw$.

Proof Thinking of u, v and w this time as being points in the plane, we can write them out in terms of their coordinates as, say, $u = (a,b)$, $v = (c,d)$, $w = (e,f)$. Then, using the definitions in 2.1.4, we find that $v + w = (c + e, d + f)$, that $u + v = (a + c, b + d)$, and that

$$u(v+w) = (a(c+e)-b(d+f), a(d+f)+b(c+e)) = (ac+ae-bd-bf, ad+af+bc+be),$$

$$uv+uw = (ac-bd, ad+bc)+(ae-bf, af+be) = (ac-bd+ae-bf, ad+be+af+bc).$$

To complete the proof, notice that the last two displays are equal. ∎

2.1.7 **Reconciliation**

Strictly speaking, the positive integer 5, the rational numbers ('fractions') $\frac{5}{1}$ and $\frac{15}{3}$, and the decimals 5.0 and 4.$\dot{9}$ are objects of different kinds. Long experience, however, has made it clear that their differences are purely representational: they do the same thing, they work the same way, they are operationally equivalent; for the purposes of purely arithmetical calculation it does not matter which you use, nor how many times you switch from one notation to another; you will obtain equivalent answers.

Likewise, since in the 'geometrical' approach the complex numbers that lie on the horizontal axis behave just like real numbers, we can feel free to view them as *being* real numbers, differently presented. That is, for each real number t, the point with coordinates $(t, 0)$ and the real number t itself can be regarded as identical and interchangeable for the purposes of purely arithmetical complex calculation. In effect, *the points on the horizontal axis are the real numbers*:

$$t = (t, 0) \quad \text{(for each real } t\text{)}.$$

(We could rerun that paragraph for the 'algebraic' approach also, but it seems scarcely necessary: for the point has already been made that the complex number $t + 0i$ and the real number t are effectively equal; *the complex numbers that have i-coefficient zero are the real numbers*.)

Back in the geometrical context, let us denote by i the point in the coordinate plane that lies one unit vertically above the origin: the point $(0, 1)$. The (still somewhat strange-looking) definition we presented concerning how points should be multiplied gives it a vitally important property:

$$i^2 =$$

$$(0, 1)(0, 1) = (0(0) - 1(1), 0(1) + 1(0)) = (-1, 0)$$

$$= -1.$$

Suddenly there is nothing semi-mythical about a square root of -1! The point one unit above the origin serves perfectly well (and so, indeed, does the point one unit below the origin).

Furthermore, for any real numbers a and b (and holding in mind that each real number *is* the point that corresponds to it on the horizontal axis) consider the simple calculation

$$a + bi$$

$$= (a, 0) + (b, 0)(0, 1) = (a, 0) + (b(0) - 0(1), b(1) + 0(0)) = (a, 0) + (0, b)$$

$$= (a, b).$$

In truth, then, there is no disagreement between the two approaches we have outlined toward what complex numbers are and how they can be conveniently annotated. The algebraically spawned symbol $a+bi$ and the geometrically sourced point-in-coordinate-plane (a, b) represent exactly the same thing—they are indeed equal subject to the understanding about points on the x-axis *being* real numbers.

A last detail that we should check before confirming reconciliation between the two presentations is that their separate definitions of addition and of multiplication are in accordance. This also is quite simple:

$$(a, b) + (c, d) = (a + c, b + d) \text{ agrees with}$$

$$a + bi + (c + di) = a + c + (b + d)i \text{ because}$$

$$a + c + (b + d)i = (a + c, b + d);$$

$$(a, b)(c, d) = (ac - bd, ad + bc) \text{ agrees with}$$

$$(a + bi)(c + di) = ac + adi + bci + bdi^2 = ac - bd + (ad + bc)i \text{ because}$$

$$(ac - bd, ad + bc) = ac - bd + (ad + bc)i.$$

Please don't think of this as just an exercise in developing decent notation. It is that, of course, but it lays the foundation for much more. The cross-fertilisation between algebraic and geometrical thinking that it draws to our attention is a recurrent theme throughout complex analysis, and a very valuable one. Even at this early stage you can see the first approach reassuring us that while manipulating complex numbers we can trust all the common-sense algebra we ever learned, provided only that we keep remembering to replace i^2 by -1... while at the same time the second approach is giving us as tangible a square root of minus one as we could have hoped for, together with indications of geometrical language such as the status of points/numbers as being on or above or below a line, or at certain distances from other objects. We shall see later that being able to locate numbers as being inside or outside various curves is an extraordinarily powerful notion in evaluating seemingly difficult integrals, and one that would be much more difficult to express or to recognise in purely algebraic terms. For the present, however, this is getting too far ahead of the game.

2.2 Navigating around the complex plane

Whenever we are thinking of complex numbers as points in the coordinate plane, we call it the *complex plane* and refer to its horizontal axis as *the real axis* rather than the *x*-axis; the vertical axis is—regrettably—often called *the imaginary axis* (although there is, of course, absolutely nothing imaginary about it).

2.2.1 Definitions   For each complex number $z = a + bi$ (where a and b are real), we call

1. a the *real part of z*, written briefly as $\mathrm{Re}(z)$,
2. b the *imaginary part of z*, written briefly as $\mathrm{Im}(z)$,
3. $a - bi$ the *conjugate of z*, written briefly as $\bar{z}$ and
4. $\sqrt{a^2 + b^2}$ the *modulus of z*, written briefly as $|z|$.

It may be worth stressing that there is also nothing imaginary about $\mathrm{Im}(z)$: it is merely a real number. Avoid making the common mistake of thinking that the imaginary part of $a + bi$ is bi!

There are simple and obvious ways to present these notions in geometrical language in the complex plane: the real numbers $\mathrm{Re}(z)$ and $\mathrm{Im}(z)$ are the coordinates of z on the horizontal and vertical axes, the complex number $\bar{z}$ is the reflection of z in the real axis, and $|z|$ (also a real number) is the distance from the origin to z.

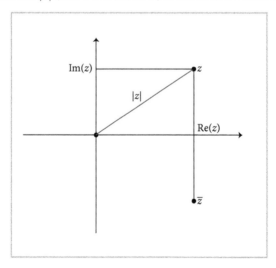

Another way to identify where in the coordinate plane (the complex plane) a point (a number) lies is to specify how far it is from the origin and in what direction: that is, to state its *polar coordinates*, the numbers r and θ indicated in this diagram:

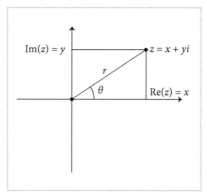

Evidently r is nothing but the modulus $|z|$ as previously defined, and θ is called the *argument* of z, written arg z. There is some lack of clarity about the angle θ that the diagram on its own does not clear up: we must measure argument angles from the positive direction of the horizontal axis, counting anticlockwise angles as positive and clockwise angles as negative. The remaining ambiguity—that angles differing by a whole number of times 2π radians (or 360 degrees) result in exactly the same point—is in most cases unimportant, but it can be resolved whenever necessary by deciding to use only angles θ that satisfy $-\pi < \theta \leq \pi$; this choice of θ is called the *principal value of the argument*. (Some textbooks use the notation Arg z to indicate this principal value.)

Whether or not the argument angle is restricted to be its principal value, the following connections between the Cartesian and the polar coordinates of a complex number should be visible from the diagram:

2.2.2 Proposition   If r and θ are the polar coordinates of the non-zero[3] complex number $z = x + yi$ then

- $x = r\cos\theta$,
- $y = r\sin\theta$,
- $r = |z| = \sqrt{x^2 + y^2}$,
- $\theta =$ an angle whose cosine is $\dfrac{x}{r}$ and whose sine is $\dfrac{y}{r}$.

The fourth of these is rather more complicated than we would all like it to be. It is tempting to use the inverse trig functions here, but we must keep in mind that arcsin only outputs values in the range $\left[-\dfrac{\pi}{2}, \dfrac{\pi}{2}\right]$ and arccos only values in $[0, \pi]$, whereas θ must at least be allowed to range over $(-\pi, \pi]$. To highlight the dangers of relying thoughtlessly on inverse trig functions, consider the question: what is the argument of $-3 - 3i$? Since in this case $r = \sqrt{(-3)^2 + (-3)^2} = 3\sqrt{2}$, if we invoke the arcsin $\left(\dfrac{y}{r}\right)$ formula we get arcsin $\left(-\dfrac{1}{\sqrt{2}}\right) = -\dfrac{\pi}{4}$, which is clearly

---

[3] Excluding 0 is little more than a way to avoid writing x/r or y/r in the one case where $r = 0$. Also the argument angle of z is actually undefined when $z = 0$.

wrong: a simple sketch diagram shows that $-3 - 3i$ actually has argument $-\dfrac{3\pi}{4}$ (or $\dfrac{5\pi}{4}$ if you prefer to work with positive angles).

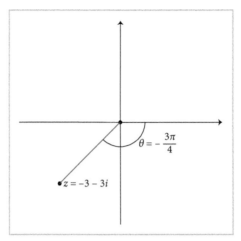

Equally, if we invoke the arccos $\left(\dfrac{x}{r}\right)$ formula, it yields $\theta = \arccos\left(-\dfrac{1}{\sqrt{2}}\right) = \dfrac{3\pi}{4}$, which is just as wrong. Thus we see the need to identify an angle whose sine and cosine are both as required. A quicker alternative is to calculate either arccos $\left(\dfrac{|x|}{r}\right)$ or arcsin $\left(\dfrac{|y|}{r}\right)$ or even arctan $\left(\dfrac{|y|}{|x|}\right)$ to find the *key angle* (that is, the acute angle between the line $0z$ and the horizontal axis) and then inspect the signs of x and of y to decide which quadrant the point z lies in, after which the argument angle should be obvious.

Here are several easily checked equalities that concern the conjugate and the modulus of general complex numbers w and z:

2.2.3 Proposition

- $z + \bar{z} = 2\,\mathrm{Re}(z)$
- $z - \bar{z} = 2i\,\mathrm{Im}(z)$
- $\overline{w + z} = \bar{w} + \bar{z}$
- $\overline{wz} = \bar{w}\,\bar{z}$
- $\overline{\left(\dfrac{w}{z}\right)} = \dfrac{\bar{w}}{\bar{z}}$ provided that $z \neq 0$
- $\bar{\bar{z}} = z$
- $|wz| = |w||z|$
- $\left|\dfrac{w}{z}\right| = \dfrac{|w|}{|z|}$ provided that $z \neq 0$

- $|\bar{z}| = |z|$
- $z\bar{z} = |z|^2$
- $\dfrac{w}{z} = \dfrac{w\bar{z}}{|z|^2}$ provided that $z \neq 0$

Notice that the last of these shows how we can easily divide one complex number by another: we 'simplify' the fraction by multiplying top and bottom by the conjugate of the bottom line. This helps because the new bottom line is purely a real number, so the problem of 'division by a complex number' has simply gone away! For instance, how can we divide $3 - 8i$ by $2 + 5i$? Like this:

$$\frac{3 - 8i}{2 + 5i} = \frac{(3 - 8i)(2 - 5i)}{(2 + 5i)(2 - 5i)} = \frac{6 - 40 - 16i - 15i}{4 + 10i - 10i + 25} = \frac{-34 - 31i}{29} = -\frac{34}{29} - \frac{31}{29}i.$$

The next two results are very important and a lot less obvious than the previous ones, so we shall set out detailed verifications for them.

2.2.4 Lemma: the triangle inequality   For any complex numbers w and z,

$$|w + z| \leq |w| + |z|.$$

Proof   We use several of the pieces of 2.2.3 together with two small observations:

1. the real part of a complex number cannot be bigger than the modulus,
2. $\overline{(z\bar{w})} = \bar{z}\,\overline{\bar{w}} = \bar{z}w.$

Now we proceed as follows:

$$\begin{aligned}
|w + z|^2 &= (w + z)\overline{(w + z)} \\
&= (w + z)(\bar{w} + \bar{z}) \\
&= w\bar{w} + z\bar{w} + \bar{z}w + z\bar{z} \\
&= |w|^2 + z\bar{w} + \overline{(z\bar{w})} + |z|^2 \\
&= |w|^2 + 2\,\mathrm{Re}(z\bar{w}) + |z|^2 \\
&\leq |w|^2 + 2|z\bar{w}| + |z|^2 \\
&= |w|^2 + 2|z||\bar{w}| + |z|^2 \\
&= |w|^2 + 2|z||w| + |z|^2 \\
&= (|w| + |z|)^2
\end{aligned}$$

and lastly take square roots (remembering that a modulus cannot be negative so no plus-or-minus ambiguity can arise).   ∎

2.2.5 Lemma: the reverse triangle inequality, a.k.a. the inverse triangle inequality

For any complex numbers w and z,

$$|w - z| \geq |\,|w| - |z|\,|.$$

Proof   Start by applying the triangle inequality to the fact that $(w - z) + z = w$, and we get

$$|w| = |(w - z) + z| \leq |w - z| + |z| \text{ and therefore}$$

$$|w| - |z| \leq |w - z|.$$

Swopping w and z and re-running this, we get

$$|z| - |w| \leq |z - w|.$$

However, the right-hand sides of these two displays are equal since

$$|z - w| = |(-1)(w - z)| = |-1||w - z| = 1|w - z| = |w - z|.$$

Also, since $|w| - |z|$ is a real number, its modulus is either $|w| - |z|$ or $(-1)(|w| - |z|) = |z| - |w|$. The above shows that, whichever one of these two it is, it must be $\leq |w - z|$.   ∎

In order to get a geometrical view of what these two inequalities are saying, we need to return to the idea of distances in the complex plane.

2.2.6 Lemma   The distance between two complex numbers w and z is $|w - z|$.

Proof

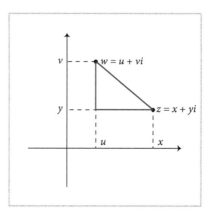

The distance is the length of the hypotenuse of the right-angled triangle indicated in the blue. The length of the horizontal side, which is either $x - u$ or $u - x$ depending on whether z lies to the right or to the left of w, is given by $\pm(u - x)$; likewise, the length of the vertical side is $\pm(v - y)$. Now Pythagoras assures us that the distance from w to z is

$$\sqrt{(u - x)^2 + (v - y)^2}.$$

Since $w - z = (u + vi) - (x + yi) = (u - x) + (v - y)i$, the displayed formula is just the definition of $|w - z|$. ∎

2.2.7 Notes

1. Now we can get a geometrical take on what the triangle inequality asserts and, indeed, where its name comes from. In the triangle OBC whose vertices are $0, z$ and $w + z$ respectively

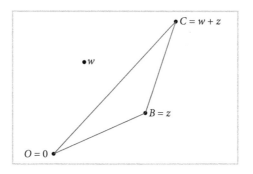

the lengths of the sides OB, BC, CO are, according to 2.2.6, $|z|$, $|(w + z) - z| = |w|$ and $|w + z|$. What 2.2.4 is asserting is that OC cannot be longer than the total length of OB and BC—in other words, the elementary geometrical insight that one side of a triangle cannot exceed in length the other two sides combined.

2. The *inverse* triangle inequality is a bit less routine to interpret, but we can illustrate it by the following diagram.

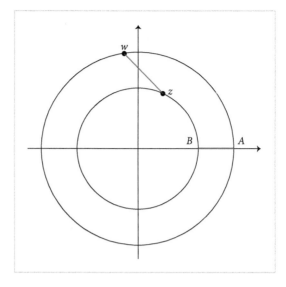

Here, w and z are two more-or-less randomly chosen complex numbers, so the (blue) line joining them has length $|w-z|$. We have also drawn the circles centre $(0,0)$ on which they lie (and whose radii are therefore $|w|$ and $|z|$). The points A and B at which these circles cross the positive half of the real axis are the real numbers $|w|$ and $|z|$, so the (red) line AB has length $|w|-|z|$ if w is further from the origin than z is, and $|z|-|w|$ otherwise. In other words, its length in both cases is $||w|-|z||$. The assertion of 2.2.5, that the blue line cannot be shorter than the red line, is merely the geometrical insight that when two circles are concentric, then the shortest possible distance from a point on one to any point on the other is the difference between their radii.

3. Looking back at the first diagram in paragraph 2.1.5, we see that each pair of opposite sides in the four-sided figure with vertices at (i) the origin, (ii) the two given complex numbers and (iii) their sum have equal length (or equally well, we can see that opposite sides have equal gradients) and so the figure is a parallelogram. The same geometrical configuration is easy to check for arbitrary points w and z in the complex plane, that is, the sum $w+z$ can be visualised as the fourth vertex of the parallelogram having the lines 0-to-w and 0-to-z as adjacent sides.

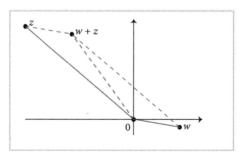

This imagery is likely to be familiar to many readers who have worked with adding vector quantities in two dimensions.

4. Finally for this section, we shall use distance measurement to interpret complex multiplication geometrically. Consider, for complex numbers w and z, the diagram

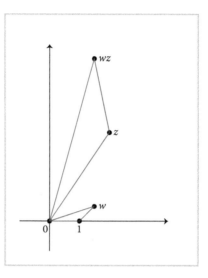

in which we have highlighted the (blue) triangle with vertices $0, 1$ and w, and the (red) triangle with vertices $0, z$ and wz. Using 2.2.6, the lengths of the sides of the first are

$$1, |w - 1|, |w|$$

and the lengths of the sides of the second are

$$|z|, |wz - z| = |(w - 1)z| = |w - 1||z|, |wz| = |w||z|.$$

So the second triangle is a scaled version of the first, with a magnification factor of $|z|$. They are therefore similar triangles, and possess equal angles at corresponding vertices. Now (continuing with the convention *anticlockwise = positive, clockwise = negative*) the red triangle's angle measured from the line 0-to-z to the line 0-to-wz is the difference $\arg(wz) - \arg(z)$

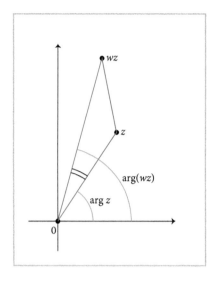

and the corresponding angle in the blue triangle is arg w. The fact that they are equal gives us

$$\arg(wz) = \arg w + \arg z$$

(and we recall the previously noted $|wz| = |w||z|$).

2.2.8 Summary   When using polar coordinates, to multiply complex numbers we must

- Multiply the Moduli, but
- Add the Arguments.

2.2.9 Alert

The one respect in which the basic arithmetic and algebra of real and of complex numbers differ is that of *order*. For each two real numbers x and y it is true that $x > y$ (x is greater than y) or that $x < y$ (x is less than y) or that they are equal, and this is a highly useful notion in many scenarios: for instance, if you need to prove that two (real) numbers must be equal, then it is good enough to show that neither can be greater than the other. No such facility is valid for complex numbers, for *there is no way to define an ordering upon* $\mathbb{C}$ *that connects sensibly with its arithmetical structure.*

As an illustration of the issue here, suppose for the sake of argument that we did have an order on $\mathbb{C}$ that behaved 'sensibly'. Consider then the question as to whether i was greater than 0 (that is, i is positive) or less than 0 (that is, i is negative). Whichever way you decide, i times i will be either positive times positive or negative times negative, and therefore positive in both cases, that is, -1

turns out to be positive. If that is not enough discouragement already, proceed to notice that $1 = (-1)(-1)$ equals positive times positive, therefore positive, and that $0 = 1 + (-1)$ equals positive plus positive and therefore positive. By now we are deep enough into contradiction to show that i cannot be classed as either greater than 0 or less than 0.

2.3 Sequences and series of complex numbers

2.3.1 Definitions   A *complex sequence* is an unending succession $(z_n)_{n \in \mathbb{N}}$ of complex numbers. If also w is a particular complex number, we say that $(z_n)_{n \in \mathbb{N}}$ *converges* to the *limit* w if, for each $\varepsilon > 0$, there is a positive integer n_ε such that

$$n \geq n_\varepsilon \text{ implies that } |z_n - w| < \varepsilon.$$

Complex sequences that possess limits are called *convergent*, and those that do not are called *divergent*. *Subsequences* of complex sequences are defined and annotated just as in the case of real sequences. Also, a complex sequence $(z_n)_{n \in \mathbb{N}}$ is called *bounded* if there is a positive real constant M such that $|z_n| < M$ for every $n \in \mathbb{N}$. (Be careful not to write or to think $-M < z_n < M$ at that point!)

2.3.2 Exercise

1. Given a complex number z such that $|z| < 1$, verify from the definition that $z^n \to 0$ as $n \to \infty$.

2. Given a complex number z such that $|z| > 1$, show that the sequence $(z^n)_{n \in \mathbb{N}}$ is divergent. (You may want to consider Proposition 2.3.4.)

Since each complex number z can be decomposed into terms of its real and imaginary parts as $z = \mathrm{Re}(z) + \mathrm{Im}(z)i$, it is reasonable to ask whether the convergence of a complex sequence can be usefully described by looking at the two real sequences comprising the real and imaginary parts of its terms. The easily confirmed answer is yes, although this is usually not the most efficient way to proceed.

2.3.3 Proposition   Let $w = u + iv$ and, for each $n \in \mathbb{N}$, $z_n = x_n + iy_n$ where u, v and each x_n and each y_n are real.[4] Then $z_n \to w$ if and only if both $x_n \to u$ and $y_n \to v$.

---

[4] Writing $x_n + iy_n$ in preference to $x_n + y_n i$ makes for slightly greater ease of reading, and saves the eye from potential confusion with 'n to the power of i'! For similar reasons, $\cos \theta + i \sin \theta$ is generally preferred to $\cos \theta + \sin \theta i$.

Proof   Notice that, for any complex numbers z and w, we have $|\operatorname{Re}(z) - \operatorname{Re}(w)| \leq |z - w|$ and $|\operatorname{Im}(z) - \operatorname{Im}(w)| \leq |z - w|$. Therefore, once we have forced $|z_n - w| < \varepsilon$, we immediately get both $|\operatorname{Re}(z_n) - \operatorname{Re}(w)| < \varepsilon$ and $|\operatorname{Im}(z_n) - \operatorname{Im}(w)| < \varepsilon$.

Conversely, note first that the triangle inequality gives

$$|z - w| \leq |\operatorname{Re}(z) - \operatorname{Re}(w)| + |\operatorname{Im}(z) - \operatorname{Im}(w)|$$

in general. Therefore, once we can guarantee that both $|\operatorname{Re}(z_n) - \operatorname{Re}(w)| < \varepsilon/2$ and $|\operatorname{Im}(z_n) - \operatorname{Im}(w)| < \varepsilon/2$, it follows that $|z_n - w| < \varepsilon$.

Hence the result.   ∎

The next group of results can be established by almost exactly the same arguments as served for their 'real' counterparts.

2.3.4 Proposition

1. Every subsequence of a convergent complex sequence is convergent, and to the same limit as the parent sequence.

2. A convergent complex sequence must be bounded.

3. **Bolzano-Weierstrass**: every bounded complex sequence possesses a convergent subsequence.

4. **Algebra of limits for complex sequences**: the limits of sums, products, quotients and so on of convergent complex sequences are exactly what common sense expects them to be; for example, if (as $n \to \infty$) $z_n \to L_1$ and $w_n \to L_2$, then $z_n w_n \to L_1 L_2$.

There is a great deal that could be said about the basic theory of *series of complex numbers* but, in practice, it is hardly worth saying: because, with a handful of exceptions, it is identical to the basic theory of real series. It will be enough to flag up a few key points:

- A complex series *converges* or *diverges* according as the sequence of its partial sums does.
- A complex series $\sum z_n$ is *absolutely convergent* if $\sum |z_n|$ is convergent.
- An absolutely convergent complex series is convergent, and will still converge to the same sum if you rearrange its terms in any fashion whatsoever.
- A complex power series $\sum c_n(z - a)^n$ will generally have a *radius of convergence* (let us denote it by r) such that the series converges absolutely at every point that lies inside the circle of centre a and radius r (its *circle of convergence*) but diverges at every point that lies outside that circle[5]: in other words, $\sum c_n(z - a)^n$

---

[5] The term *radius of convergence* usually puzzles learners when they meet it first in real analysis, but it makes perfectly good sense once you see it in terms of complex numbers.

will converge absolutely for every z such that $|z - a| < r$, but diverge for every z such that $|z - a| > r$. The exceptional types are are (i) certain such series converge *only* at $z = a$ and are said to have radius of convergence 0, while (ii) certain important such series converge absolutely at *every* point of the complex plane, and for these the radius of convergence is said to be infinite.

- *Term-by-term differentiation* works for complex power series just as it does for real ones: that is, if $\sum c_n(z - a)^n$ has non-zero radius of convergence r and $f(z)$ denotes its sum at the typical point z inside the circle of convergence, then f is differentiable[6] there, and $f'(z) = \sum n c_n(z-a)^{n-1}$. It may be worth restating that in order to bring to light just how natural and inevitable it looks, notwithstanding how difficult it is to prove it properly:

$$\text{if } f(z) = c_0 + c_1(z - a) + c_2(z - a)^2 + c_3(z - a)^3 + c_4(z - a)^4 + \cdots$$

$$\text{then } f'(z) = c_1 + 2c_2(z - a) + 3c_3(z - a)^2 + 4c_4(z - a)^3 + \cdots$$

where, for every point z inside the circle of convergence, both series are not only convergent, but *absolutely convergent*.

- The comparison tests cannot be used directly on complex series since conditions such as $0 \le a_n \le b_n$ make no sense at all once a_n and b_n are complex, but they can of course be used on the associated real series $\sum |a_n|$ and $\sum |b_n|$.

- Likewise, the ratio test and the nth root test cannot be used directly on a complex series $\sum z_n$ since the limit of the growth rate or of the nth root of the nth term (if it exists) is going to be a complex number, and so it will be meaningless to ask whether it is greater or smaller than 1. On the other hand, they can often be applied to the associated real series $\sum |z_n|$ (and if it converges, then the original series is absolutely convergent, and therefore convergent).

2.3.5 Example

Show that, for each complex number z such that $|z| < 1$, the *geometric series*

$$1 + z + z^2 + z^3 + z^4 + \cdots$$

is absolutely convergent, and that its sum is $\dfrac{1}{1 - z}$.

Solution From the original series $\sum_0^\infty z^n$ we obtain the modulussed series $\sum_0^\infty |z|^n$, a real series whose growth rate is (constantly) $|z|$ which is less than 1. By the ratio test, the modulussed series is convergent; that is, the original series is absolutely convergent.

---

[6] Strictly speaking, we have not yet defined differentiation for complex expressions—this will be done in paragraph 4.2.22—but for the moment be reassured that its elementary aspects work in exactly the same way as the differentiation of real expressions; for instance, the derivatives of $(z + a)^n$ and e^{az} are $n(z + a)^{n-1}$ and ae^{az} for $n \in \mathbb{N}$ and constant a.

Also the nth partial sum $S_n = 1 + z + z^2 + z^3 + z^4 + \cdots + z^{n-1}$ satisfies the equation $(1 - z)S_n = 1 - z^n$ (because everything else on the left-hand side cancels), and therefore

$$S_n = \frac{1 - z^n}{1 - z}$$

for each n. Letting $n \to \infty$ and observing that $z^n \to 0$, we get the sum (to infinity) of the series to be as predicted.

One more 'good behaviour' aspect of absolutely convergent series (real or complex) is that they can be multiplied together in a very natural way that respects their convergence and their limits. We shall use this solely in the case of power series so we shall state it only in that context where, indeed, it looks like pure common sense. (Nevertheless, once again, it is quite a demanding challenge to assemble a logically watertight proof.)

2.3.6 **Theorem and definition**   Suppose that two absolutely convergent series converge to functions $f(z)$ and $g(z)$ inside a circle centre 0, thus:

$$a_0 + a_1 z + a_2 z^2 + a_3 z^3 + \cdots + a_n z^n + \cdots = f(z),$$

$$b_0 + b_1 z + b_2 z^2 + b_3 z^3 + \cdots + b_n z^n + \cdots = g(z).$$

Then their *Cauchy product*, the series

$$a_0 b_0 + (a_0 b_1 + a_1 b_0)z + (a_0 b_2 + a_1 b_1 + a_2 b_0)z^2 + (a_0 b_3 + a_1 b_2 + a_2 b_1 + a_3 b_0)z^3 + \cdots$$

also converges there absolutely, and its sum is $f(z)g(z)$.

The same *pattern* of multiplication works equally well if the power series use different variables, that is,

$$a_0 + a_1 z + a_2 z^2 + a_3 z^3 + \cdots + a_n z^n + \cdots = f(z) \text{ and}$$

$$b_0 + b_1 w + b_2 w^2 + b_3 w^3 + \cdots + b_n w^n + \cdots = g(w)$$

in which scenario the product series will converge to $f(z)g(w)$.

2.3.7 **Exercise**   Take (from paragraph 2.3.5) the power series for $\dfrac{1}{1 - z}$ and the power series for $\dfrac{1}{1 + z}$, and form their Cauchy product. Check that its sum is the product of $\dfrac{1}{1 - z}$ by $\dfrac{1}{1 + z}$ (that is, $\dfrac{1}{1 - z^2}$) as predicted by 2.3.6.

2.3.8 Exercise   The complex exponential, sine and cosine functions are defined by the series

$$e^z = 1 + z + \frac{z^2}{2!} + \frac{z^3}{3!} + \frac{z^4}{4!} + \frac{z^5}{5!} + \frac{z^6}{6!} + \frac{z^7}{7!} + \cdots$$

$$\sin z = z - \frac{z^3}{3!} + \frac{z^5}{5!} - \frac{z^7}{7!} + \cdots$$

$$\cos z = 1 - \frac{z^2}{2!} + \frac{z^4}{4!} - \frac{z^6}{6!} + \cdots .$$

Show that all three series are absolutely convergent for every $z \in \mathbb{C}$. (Notice that, if we replace z by a real number t, these become the standard power series expressions for e^t, $\sin t$ and $\cos t$. So the effect of the definitions is to extend these familiar real functions from the real axis to the whole complex plane.)

Partial solution   For the first series $\sum \frac{z^n}{n!}$ (and assuming $z \neq 0$), apply the ratio test to the modulussed series $\sum \left| \frac{z^n}{n!} \right| = \sum \frac{|z|^n}{n!}$ and see that the limit of the growth rate is 0. Therefore $\sum \frac{z^n}{n!}$ is absolutely convergent.

2.4 Powers and roots: de Moivre's theorem

Now that we have power series descriptions of basic (complex) exponential and trigonometric functions, look at the effect of replacing z in the first of them by $i\theta$ where θ is real[7]:

$$e^{i\theta} = 1 + i\theta + \frac{(-1)\theta^2}{2!} + \frac{(-i)\theta^3}{3!} + \frac{(+1)\theta^4}{4!} + \frac{(i)\theta^5}{5!} + \frac{(-1)\theta^6}{6!} + \frac{(-i)\theta^7}{7!} + \cdots$$

$$= \left(1 - \frac{\theta^2}{2!} + \frac{\theta^4}{4!} - \frac{\theta^6}{6!} + \cdots \right) + i \left(\theta - \frac{\theta^3}{3!} + \frac{\theta^5}{5!} - \frac{\theta^7}{7!} + \cdots \right)$$

$$= \cos \theta + i \sin \theta.$$

This powerful and succinct little connection between what might appear to be two quite different types of elementary function is known as *Euler's formula*.

2.4.1 Euler's formula   For any real θ, $e^{i\theta} = \cos \theta + i \sin \theta$.

---

[7] By the way, the infinite amount of rearrangement taking place here illustrates the power and usefulness of results such as 1.4.8 and of its 'complex clone' that we mentioned in the third bullet point between 2.3.4 and 2.3.5.

2.4.2 Note   As a first consequence of this insight, think again how we can best write down a complex number $z = x + iy$ in terms of its polar coordinates r (the modulus) and θ (the argument angle). We noted some time ago that, in this scenario, $x = r\cos\theta$ and $y = r\sin\theta$, so that $z = r\cos\theta + ir\sin\theta = r(\cos\theta + i\sin\theta)$. Now Euler allows us to compress this into a very compact form: $re^{i\theta}$.

- **Notation**: the complex number of modulus r and argument θ is $re^{i\theta}$.

As often happens in mathematics, good notation is a facilitator. For now look how completely natural the multiplication of complex numbers in polar coordinates appears:

$$\left(re^{i\theta}\right)\left(se^{i\phi}\right) = (rs)e^{i(\theta+\phi)}$$

(that's how the exponential function behaves[8]) which visibly has modulus rs and argument $\theta + \phi$. Effectively we have 'rediscovered' the conclusion of 2.2.8, but it now appears as a simple consequence of the index laws.

Continuing, when we calculate the square, the cube and the fourth power of a typical $z = re^{i\theta}$:

$$z^2 = \left(re^{i\theta}\right)^2 = \left(re^{i\theta}\right)\left(re^{i\theta}\right) = r^2 e^{2i\theta}$$
$$z^3 = z^2 z = \left(r^2 e^{2i\theta}\right)\left(re^{i\theta}\right) = r^3 e^{3i\theta}$$
$$z^4 = z^3 z = \left(r^3 e^{3i\theta}\right)\left(re^{i\theta}\right) = r^4 e^{4i\theta}$$

the emerging pattern—almost too obvious as to require any formal proof—is a key result known as *de Moivre's theorem*. (Note that integer powers of complex numbers are defined exactly as for real numbers: z^1 means z, $z^{n+1} = z^n z$ for $n \in \mathbb{N}$ and (provided that $z \neq 0$) $z^0 = 1$ and $z^{-n} = 1/z^n$ for $n \in \mathbb{N}$.)

2.4.3 de Moivre's theorem   For any (non-zero) complex number $z = re^{i\theta}$ and any integer m, we have

$$z^m = r^m e^{im\theta}.$$

Proof anyway

1. When m is positive, this result can be established from the preceding roughwork as a very easy exercise in induction, which we leave for the interested reader.

---

[8] As you know, $e^a e^b = e^{a+b}$ for any real numbers a and b; it is routine to check that this rule also holds for complex indices: see Exercise 17 at the end of this chapter.

2. When $m = 0$, all that is asserted is that

$$z^0 = r^0 e^{0i} = r^0(\cos 0 + i \sin 0)$$

which is little more than the algebraic convention that 'anything to the power of zero' is 1.

3. In the remaining case, m is a negative integer, that is, $m = -n$ where n is a positive integer. From case (a)—already established—we know that $z^n = r^n e^{in\theta}$. Also z^m means $\dfrac{1}{z^n}$. Yet

$$\left(r^m e^{im\theta}\right)\left(r^n e^{in\theta}\right) = r^{m+n} e^{i(m+n)\theta} = r^0 e^{0i} = 1$$

from which we see that

$$\frac{1}{r^n e^{in\theta}} = r^m e^{im\theta}.$$

Putting the fragments together, we confirm that

$$z^m = \frac{1}{z^n} = \frac{1}{r^n e^{in\theta}} = r^m e^{im\theta}$$

in the negative-m case also. ∎

2.4.4 Aside

Although it is rather peripheral to the aim of the present text, we ought to point out that de Moivre gives one of the first indications that 'complex' calculations can have useful 'real' consequences. How, for instance, could we express $\cos(7\theta)$ in terms of powers of $\cos \theta$? The traditional 'real' method is to break it down into

$$\cos(3\theta + 4\theta) = \cos(3\theta)\cos(4\theta) - \sin(3\theta)\sin(4\theta)$$

and then further decompose each of these four components into terms of lower multiples of θ. It works, but it takes significant time and patience. Alternatively, de Moivre offers the observation that $\cos(7\theta)$ is the real part of $e^{7i\theta} = (\cos\theta + i\sin\theta)^7$, which we could immediately expand via the binomial theorem and then simply *pick out* what the real part is.

2.4.5 **Example**   Express $\cos(7\theta)$ in terms of $\cos\theta$.

Solution   For convenience let $c = \cos\theta, s = \sin\theta$. By the binomial theorem,

$$(c + is)^7 = c^7 + 7c^6(is) + 21c^5(is)^2 + 35c^4(is)^3 + 35c^3(is)^4 + 21c^2(is)^5 + 7c(is)^6 + (is)^7$$

$$= c^7 + 7ic^6s - 21c^5s^2 - 35ic^4s^3 + 35c^3s^4 + 21ic^2s^5 - 7cs^6 - is^7$$

whose real part[9] is

$$c^7 - 21c^5s^2 + 35c^3s^4 - 7cs^6$$

$$= c^7 - 21c^5(1 - c^2) + 35c^3(1 - 2c^2 + c^4) - 7c(1 - 3c^2 + 3c^4 - c^6)$$

$$= 64c^7 - 112c^5 + 56c^3 - 7c.$$

Since also $(c + is)^7 = \left(e^{i\theta}\right)^7 = e^{7i\theta} = \cos(7\theta) + i\sin(7\theta)$ via de Moivre, we get

$$\cos(7\theta) = 64\cos^7\theta - 112\cos^5\theta + 56\cos^3\theta - 7\cos\theta.$$

2.4.6 **Note**   When complex numbers are thought of as points in a Cartesian coordinate plane and written in the form $x + iy$ or (x, y) with x and y being real, there is no trace of ambiguity as to what *equals* means: $x + iy = x' + iy'$ happens if and only if both $x = x'$ and $y = y'$. With polar coordinates, on the other hand, ambiguity does exist concerning the argument angle: for $re^{i\theta}$ to equal $se^{i\phi}$ we need them to be the same distance from the origin and in the same direction, so certainly r and s must be equal; but θ and ϕ could be different: for if they were to differ by a whole number of complete revolutions—say, $2m\pi$ for some integer m—then the directions that they specify would be indistinguishable. In other terms,

$$re^{i\theta} = se^{i\phi} \text{ if and only if both } r = s \text{ and } \phi = \theta + 2m\pi \text{ for some integer } m.$$

Although this may appear to be a cumbersome way to view equality, it is exactly what we need in order to identify nth roots of complex numbers.

2.4.7 **Definition**   Suppose z is a complex number and n is a positive integer. Any complex number w for which $w^n = z$ is called an *nth root of z*. (Since it should be obvious that the only nth root of zero is zero, we shall assume that $z \neq 0$ going forward.)

2.4.8 **Proposition**   Every non-zero complex number has exactly n distinct nth roots.

Solution   We write z as $re^{i\theta}$ and w as $se^{i\phi}$. For w to be an nth root of z, what we need is that $w^n = z$, that is, that $s^n e^{in\phi} = re^{i\theta}$. From the above discussion, this says precisely that $s^n = r$ and that $n\phi = \theta + 2m\pi$ for some integer m. Equivalently, s is the positive real nth root of r, and ϕ takes the form

$$\phi = \frac{\theta + 2m\pi}{n}, \quad m = 0, \pm1, \pm2, \pm3, \pm4, \cdots .$$

[9] Remember that $s^2 = 1 - c^2, s^4 = (1 - c^2)^2$ and so on.

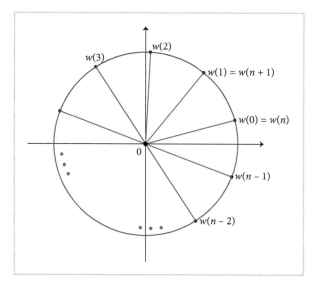

Since these numbers are uniformly spaced around the circle of centre 0 and radius $\sqrt[n]{r}$ at argument angles of $\dfrac{2\pi}{n}$ (that is, one nth of a complete revolution), exactly n of them are distinct: for instance, those given by $m = 0, 1, 2, 3, \cdots, n-1$ are all different, beyond which all other values of m give a root already accounted for.   ∎

It may be useful to restate the proposition in a manner that delivers explicitly what we find the roots in question to be:

2.4.9 Proposition modified

For non-zero $z = re^{i\theta}$ and positive integer n, z has exactly the n distinct n^{th} roots

$$w(0), w(1), w(2), \cdots, w(n-1)$$

given by

$$w(m) = \sqrt[n]{r}\, e^{i(\theta + 2m\pi)/n}, \quad m = 0, 1, 2, 3, \cdots, n-1.$$

2.5 Exercises

1. Using the 'algebraic' notation (in which a typical complex number is a symbol of the form $x + iy$) show that, for any complex numbers u, v and w, we have
   (a) $uv = vu$,
   (b) $u(v + w) = uv + uw$.

2. Using the 'geometrical' notation (in which a typical complex number is a point in the coordinate plane) show that
   (a) for any non-zero u in $\mathbb{C}$ there is a complex number u^{-1} such that $uu^{-1} = 1$,
   (b) for any u, v, w in $\mathbb{C}$ we have $u(vw) = (uv)w$.

3. Express each of the following in polar coordinates:
   (a) $-6\sqrt{3} + 6i$,
   (b) $-2 - 2i$,
   (c) $20i$,
   (d) $\dfrac{1}{5 + 5\sqrt{3}i}$.

4. Express each of the following in '$x + iy$' form:
   (a) $100e^{\pi i/3}$,
   (b) $2e^{\pi i/4} + 3e^{3\pi i/4}$,
   (c) $(1 - i)^6$,
   (d) $\dfrac{(2 + 2\sqrt{3}i)^4}{(2 - 2i)^3}$.

5. Given that $|z| = 5$, show that
   (a) $10 \le |3z + 5i| \le 20$,
   (b) $23 \le |z^2 + 1 + \sqrt{3}i| \le 27$
   and see if you can determine the maximum value that is actually achieved by the expression $|z^2 + 1 + \sqrt{3}i|$ while $|z| = 5$.

6. Given that z lies in the annulus $\{z \in \mathbb{C} : 7 \le |z| \le 8\}$ and that $w = 1 - z + z^2$, show that w must lie in the annulus $\{w \in \mathbb{C} : 40 \le |w| \le 73\}$.
   Is it possible for $|w|$ to equal 73?
   Is it possible for $|w|$ to equal 40?

7. (As stated in paragraph 2.2.3) verify that, for every pair of complex numbers w and z,
   (a) $\overline{wz} = \overline{w}\,\overline{z}$,
   (b) $|wz| = |w|\,|z|$,
   (c) $\left|\dfrac{w}{z}\right| = \dfrac{|w|}{|z|}$ provided that $z \ne 0$.

8. (Provided that neither z nor w is zero) show that $\operatorname{Im}\left(\dfrac{w}{z}\right) = 0$ if and only if $\arg w$ and $\arg z$ differ by an integer multiple of π.

9. Given that the complex number z satisfies the condition

$$|z + 1| = K|z - 1|$$

where $K > 1$ is a real constant, show that z lies on a circle in the complex plane, and determine the coordinates of the centre and the radius of this circle.

10. Extend the Bolzano-Weierstrass theorem to the complex numbers – that is, *assuming* that every bounded sequence of real numbers possesses a convergent subsequence, prove that every bounded sequence of complex numbers possesses a convergent subsequence.

11. Find the limit of each of the complex sequences whose n^{th} terms are given as follows:

(a) $\dfrac{3n^2 - 2n + 5}{n^2 + 4} + \dfrac{n^3}{n^3 - 6}i,$

(b) that n^{th} root of $12 + 5i$ whose argument is positive and as small as possible,

(c) $\left(\dfrac{5 + 7i}{9}\right)^n.$

12. (For $|z| < 1$) find the sum of each of the following series:

(a) $1 + 2z + 3z^2 + 4z^3 + \cdots + nz^{n-1} + \cdots,$

(b)

$$\sum_{k=0}^{\infty} \frac{(k+1)(k+2)(k+3)}{6} z^k.$$

13. The **hyperbolic functions** sinh and cosh are defined by the equations

$$\sinh(t) = \frac{e^t - e^{-t}}{2}, \quad \cosh(t) = \frac{e^t + e^{-t}}{2}$$

where t can be either real or complex. Show that (for real or complex t):

(a) $(\cosh t)' = \sinh t$ and $(\sinh t)' = \cosh t,$

(b) $\cosh(it) = \cos t$ and $\cos(it) = \cosh t,$

(c) $\sinh(it) = i \sin t$ and $\sin(it) = i \sinh t,$

(d) $\cosh^2 t - \sinh^2 t = 1,$

(e) $\sinh(2t) = 2 \sinh t \cosh t,$

(f) $\cosh(2t) = \cosh^2 t + \sinh^2 t.$

14. (a) Express $\cos(5\theta)$ as a polynomial in $\cos \theta$, and $\sin(7\theta)$ as a polynomial in $\sin \theta$.

(b) Determine constants a, b and c such that $\sin^4 \theta$ can be expressed (for all real θ) as $a \cos(4\theta) + b \cos(2\theta) + c$, and obtain a similar expression for $\cos^6 \theta$. (Begin by verifying that if $z = e^{i\theta}$ then $z^n + z^{-n} = 2 \cos(n\theta)$ and $z^n - z^{-n} = 2i \sin(n\theta)$ for each $n \in \mathbb{N}$.)

15. (a) If an nth-degree polynomial

$$p(t) = a_n t^n + a_{n-1} t^{n-1} + a_{n-2} t^{n-2} + \cdots + a_1 t + a_0$$

factorises completely into the form

$$p(t) = a_n (t - t_1)(t - t_2)(t - t_3) \cdots (t - t_n)$$

show that the sum of its roots $t_1 + t_2 + t_3 + \cdots + t_n$ is equal to $-\dfrac{a_{n-1}}{a_n}$.

(Here, the variable and the various coefficients and roots may be either real or complex.)

(b) Prove that if w is any non-zero complex number and n is an integer greater than or equal to 2, then the sum of all the distinct nth roots of w is zero.

(c) Deduce that $\sin(\pi/17)+\sin(2\pi/17)+\sin(3\pi/17)+\cdots+\sin(16\pi/17)$ equals 0.

(d) Evaluate $\cos(\pi/17) + \cos(2\pi/17) + \cos(3\pi/17) + \cdots + \cos(16\pi/17)$.

16. Taking the complex trigonometric functions to be defined by the power series in 2.3.8, form the (Cauchy) product of the series for $\sin z$ by the series for $\cos z$, at least as far as the term in z^7. Use this together with 2.3.6 as partial evidence for the identity $\sin(2z) = 2\sin z \cos z$.

17. Taking the complex exponential function to be defined by the power series in 2.3.8, form the (Cauchy) product of the series for e^z by the series for e^w. Use this together with 2.3.6 to establish the identity $e^z e^w = e^{z+w}$. (You will also need the binomial theorem.)

3 Background part B
Real functions and their limits

As we explained at the beginning of Chapter 1, the three background chapters (numbers 1, 3 and 5) intend only to revise briefly those aspects of basic real analysis that we shall need to use, and to encourage those readers who, through reading that overview, find themselves underprepared, to make the time to acquire enough skill and understanding to proceed confidently. An obvious way to do this is to refer to one of the texts referenced at the end of the book.

3.1 Real functions

3.1.1 Definition   A *real function* is a map (or mapping or transformation) from a subset A of $\mathbb{R}$ (the set of real numbers) to a subset B of $\mathbb{R}$. That is, it is a procedure of some kind that allows us to determine, for each individual number x in A, a single 'corresponding' number in B. The sets A and B are called the *domain* and the *codomain*. The group of symbols

$$f : A \to B$$

says that f is a real function from domain A to codomain B (and then $f(x)$ denotes the element of B that corresponds to a typical element x of A). In most of the cases that concern us, the mapping will be made specific by stating some kind of formula telling us how to calculate $f(x)$ for each relevant x. When this is so, the default domain A (unless otherwise specified) will be the set of all real x for which the formula makes sense (and delivers a real number), and the default codomain will be the whole of $\mathbb{R}$. So long as all numbers under consideration are real, we generally say simply 'function' rather than 'real function'.

For instance, the formulae

$$f(x) = ax^2 + bx + c, \quad g(x) = \sqrt{18 - 2x^2}, \quad h(x) = 4\sin(\pi(x + 0.25))$$

define three particular real functions (the general quadratic, a particular square root function, and a slight modification of the trigonometric sine function). A little thought will show that

Integration with Complex Numbers. McCluskey and McMaster, Oxford University Press.
© Brian McMaster and Aisling McCluskey (2022). DOI: 10.1093/oso/9780192846075.003.0003

$$f : \mathbb{R} \to \mathbb{R}, \quad g : [-3, 3] \to \mathbb{R} \quad \text{and} \quad h : \mathbb{R} \to \mathbb{R}.$$

In fairly simple cases such as these, it is perfectly legitimate to speak of 'the function $f(x) = ax^2 + bx + c$' or even 'the function $\sqrt{18 - 2x^2}$'.

It would also have been legitimate—had we had a reason to do this—to consider that

$$g : [-3, 3] \to [0, 3\sqrt{2}] \quad \text{and that} \quad h : \mathbb{R} \to [-4, 4].$$

A function $f : A \to B$ is called *one-to-one* if the equation $f(x) = f(y)$ is true *only* in the obvious special case where $x = y$: in less formal words, when f never assumes the same value twice or more.

We draw attention to two simple and very useful classes of function:

3.1.2 Definition

- A *polynomial* is any function of the form

$$a_0 + a_1 x + a_2 x^2 + a_3 x^3 + \cdots + a_n x^n$$

where the non-negative integer n is called its *degree* provided that $a_n \neq 0$, and the constants $a_0, a_1, \cdots, a_n$ are called its *coefficients*. The default domain for a polynomial is $\mathbb{R}$ since such a formula makes sense for any real value of x.

- A *rational function* means one polynomial divided by another: that is, it is any function of the form

$$\frac{a_0 + a_1 x + a_2 x^2 + a_3 x^3 + \cdots + a_n x^n}{b_0 + b_1 x + b_2 x^2 + b_3 x^3 + \cdots + b_m x^m}.$$

Once we identify the 'zeros' of the bottom line, that is, the elements of the set

$$S = \{ x \in \mathbb{R} : b_0 + b_1 x + b_2 x^2 + b_3 x^3 + \cdots + b_m x^m = 0 \},$$

then the default domain of the rational function is $\mathbb{R} \setminus S$ (since the formula makes sense for any real value of x *except* those that cause the bottom line to take the value zero).

3.1.3 Definition   It is sometimes possible to combine functions by 'concatenation' into composite functions. That is, if $f : A \to B$ and $g : B' \to C$ where[1] $B \subseteq B'$

---

[1] It is good practice, although not strictly necessary, to arrange that B and B' be actually equal as sets.

(ensuring that every value $f(x)$ of f lies within the domain of g) then we can define the *composite function* $g \circ f : A \rightarrow C$ by the formula

$$(g \circ f)(x) = g(f(x)) \quad \text{for each } x \text{ in } A.$$

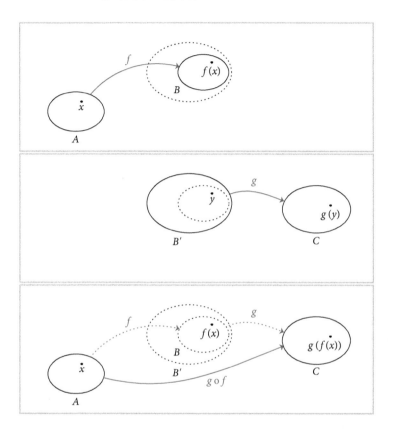

The idea extends naturally to composites of three or more functions provided only that no function is ever applied to a number that lies outside its domain.

For instance, the function that we called g in paragraph 3.1.1 can be thought of as a composite: defining, say, $\alpha : [-3, 3] \rightarrow [0, \infty)$ by $\alpha(x) = 18 - 2x^2$ and $\beta : [0, \infty) \rightarrow \mathbb{R}$ by $\beta(x) = \sqrt{x}$, the composite $\beta \circ \alpha$ is, indeed, g. In much the same way, the function referred to as h in paragraph 3.1.1 can be viewed as the composite of four functions $p(x) = x + 0.25$, $q(x) = \pi x$, $r(x) = \sin x$ and $s(x) = 4x$, so long as we make sensible and coherent choices of their domains and codomains. Decomposing a relatively complicated function—we mean, a function whose defining formula is relatively complicated—into a composite of much simpler ones turns out to be very useful in differentiation, as we shall remind ourselves in Section 3.4.

3.1.4 **Note**   The *graph* of a (real) function $f : A \to B$ is the set of all points (x, y) in the coordinate plane for which $y = f(x)$; that is, it is

$$\{(x, f(x)) : x \in A\}.$$

Full knowledge of the graph tells you everything there is to know about the function (provided that the codomain is known) and *vice versa*; indeed, from a more formal point of view, a function *is* its graph. But even informally, a sketch graph of a given function will often turn out to be a useful visual prop to our thinking as well as a way to store information about it (*Is it increasing or decreasing? Has it a biggest or a smallest value somewhere? Are there any breaks or sharp corners?*) provided always that we don't over-depend on the doubtful accuracy of back-of-the-envelope graphics.

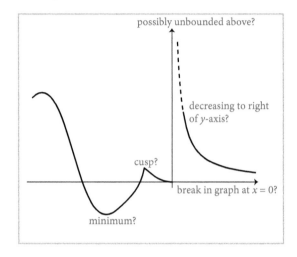

3.2 Limits of real functions

Whether or not a function f is defined at a particular point p, it often happens that the values of $f(x)$ become extremely close to a 'limit' number L as x gets closer and closer to p. The classic example of this is the behaviour of the function $f(x) = \dfrac{\sin x}{x}$ as x gets very close to 0: for although in this case '$f(0)$' makes no sense at all, values of x very close to but not exactly equal to 0 will force $f(x)$ to become as close to the number 1 as anyone can ask them to be.

Here is a sketch graph of f to aid us in imagining what happens as x gets extremely close to 0. Note that the point $(0, 1)$ which common sense 'expects' to be on this graph is missing[2]: because $f(0)$ is undefined.

---

[2] The size of the 'missing' point has been exaggerated in the sketch graph in order to draw attention to it. Likewise, so has the size of the 'replacement' points in the next two sketch graphs.

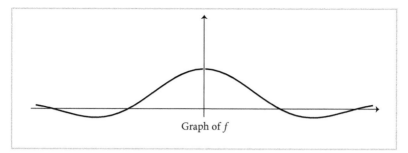

Graph of f

It is important to realise that, if we had made a decision as to the value of this function *at* 0, it would have had no bearing whatsoever on the limit notion that we are here describing. Had we, for instance, defined a slightly different function f_1 by the split-level formula

$$f_1(x) = \begin{cases} \dfrac{\sin x}{x} & \text{if } x \neq 0, \\[2mm] 2 & \text{if } x = 0 \end{cases}$$

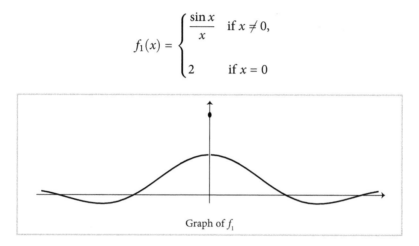

Graph of f_1

or another slightly different (but perhaps more sensible?) function f_2 using the formula

$$f_2(x) = \begin{cases} \dfrac{\sin x}{x} & \text{if } x \neq 0, \\[2mm] 1 & \text{if } x = 0 \end{cases}$$

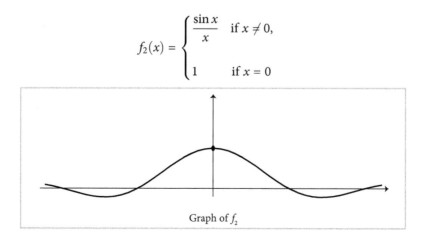

Graph of f_2

then the comment about the function's value at x becoming arbitrarily close to the limiting number 1 as x becomes extremely close to (but distinct from) 0 remains exactly as valid for f_1 and for f_2 as it was for f. *Limits of $f(x)$ as x approaches p do not care at all what—if anything—f does at the exact point p, but only what it does very close to p.*

The above discussion and illustrations are evidently not precise enough to support any serious mathematical investigation and, in fact, there are two distinct and highly useful ways to address this lack of exactness: one that involves limits of sequences (a topic that we already summarised in Chapter 1) and the other that focuses directly on the size of the error that we make when we say '$f(x)$ is approximately equal to L (when x is close enough to p)'. Here is the sequence-based one:

3.2.1 Definition: function limits via sequences

We shall say that a function $f : A \to B$ *tends* (or *converges*) to a *limit* L in $\mathbb{R}$ as x tends (or converges) to p if, for *every* sequence $(x_n)_{n\in\mathbb{N}}$ in $A \setminus \{p\}$ such that $x_n \to p$, we have $f(x_n) \to L$. In this case we write

$$\lim_{x \to p} f(x) = L$$

or, equivalently,

$$f(x) \to L \text{ as } x \to p.$$

3.2.2 Example

Use sequences to determine the limit, as $x \to 2$, of the function
$$f(x) = \frac{x^3 - 8}{x^5 - 32}.$$

Solution   The (default) domain is $(-\infty, 2) \cup (2, \infty)$ since the stated formula returns a real answer for every $x \in \mathbb{R}$ except $x = 2$.

Let (x_n) denote any sequence in the domain of f such that $x_n \to 2$. Then

$$f(x_n) = \frac{x_n^3 - 8}{x_n^5 - 32}.$$

(*Roughwork:* We cannot immediately find its limit by evaluating the limits of the top and bottom lines separately, since both are converging to zero and, of course, zero divided by zero makes no sense. However, $t^3 - 8 = (t - 2)(t^2 + 2t + 4)$ and $t^5 - 32 = (t - 2)(t^4 + 2t^3 + 4t^2 + 8t + 16)$ for every real t. So:)

$$f(x_n) = \frac{x_n^3 - 8}{x_n^5 - 32} = \frac{x_n^2 + 2x_n + 4}{x_n^4 + 2x_n^3 + 4x_n^2 + 8x_n + 16} \to \frac{4 + 4 + 4}{16 + 16 + 16 + 16 + 16} = \frac{3}{20}$$

as $n \to \infty$. (See the last sentence of 1.3.2 if this is not clear.) Hence $f(x) \to \frac{3}{20}$ as $x \to 2$. (There are, of course, alternative and more efficient ways of identifying this limit; ultimately, however, they are all built upon definitions such as this.)

Here now is the direct-error-measuring definition:

3.2.3 Definition: function limits via 'epsilontics'   We shall say that a function $f : A \to B$ *tends* (or *converges*) to a *limit L* in $\mathbb{R}$ as x tends (or converges) to p if, for *every* positive real number ε (thought of as the permissible tolerance of error for some particular application), we can find a positive number δ (thought of as *how close to p* we need to be in order to guarantee that the actual error shall be smaller than the tolerance) such that

for every x in A such that $0 < |x - p| < \delta$, we find that $|f(x) - L| < \varepsilon$.

In this case we (again) write $\lim_{x \to p} f(x) = L$ or, equivalently, $f(x) \to L$ as $x \to p$.

3.2.4 Example   Considering the function $f(x) = \sqrt[3]{x} \sin\left(\frac{1}{x}\right)$ for $x \neq 0$, show via an epsilontic argument that $f(x) \to 0$ as $x \to 0$.

Solution   Provided always that $x \neq 0$,

$$|f(x) - 0| = \left| \sqrt[3]{x} \sin\left(\frac{1}{x}\right) \right| = \left| \sqrt[3]{x} \right| \left| \sin\left(\frac{1}{x}\right) \right| \leq \left| \sqrt[3]{x} \right| .$$

Given any $\varepsilon > 0$, define δ to be ε^3 (which, indeed, is also greater than zero). Then $0 < |x - 0| < \delta$ implies that $|f(x) - 0| \leq |\sqrt[3]{x}| = \sqrt[3]{|x|} < \sqrt[3]{\varepsilon^3} = \varepsilon$. Hence the result.

3.2.5 Notes

- Of course the phrases in parentheses in 3.2.3 are not strictly part of the definition, and do not normally appear in more formal texts; they are included here merely to try to supply some insight as to what the definition is seeking to express: that one can ensure that the error we make in saying '$f(x)$ is a good approximation to L' can be made smaller than absolutely any predetermined tolerance, merely by working within some clearly identified small distance from p... but definitely not allowing x to equal the rogue point p exactly, and (of course) staying within the domain of the function since otherwise $f(x)$ would not even make sense.

- There are several points that should be stressed:

1. Despite looking rather different, definitions 3.2.1 and 3.2.3 are entirely equivalent: if either one is satisfied, then so is the other;
2. therefore you are at liberty to choose either of them whenever you have to work through a function limit question, and it will make no difference logically which you opt for;
3. but in practice, some problems (rather a lot of them, actually) are found to work out more quickly and more easily using 3.2.1, although for a substantial

minority (including quite a few of the more difficult ones) 3.2.3 is better suited and more efficient; therefore you do need to know both definitions.

4. Not all functions possess limits at all relevant points. For instance, a function whose graph resembles one of the following sketches will fail to have any limit at the 'break point' in mid-diagram.

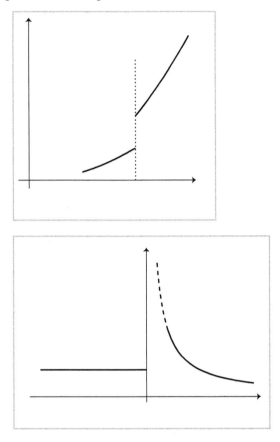

5. For the limit as $x \to p$ of a function f to make sense, it is necessary that the domain of f include points that are arbitrarily close to p (since otherwise the definitions would be trying to discuss either sequences or x-values that did not exist). For instance, it would be meaningless to try to investigate the limit of the real function $\log x$ as $x \to -3$, or the real function $\sqrt{x}$ as $x \to -4$, or the real function $\arcsin x$ as $x \to 2$ since, in each case, the relevant function has absolutely no domain points within one unit (or more) from the point at which attention focused.

6. Despite the (relative) difficulty most learners experience in mastering Definitions 3.2.1 and 3.2.3, many of the key results that can be proved from them seem very natural and predictable. The following three (which we shall have occasion to use) should illustrate this point.

3.2.6 Theorem: an algebra of limits for real functions   Suppose that f and g are two functions, that K is a real constant, and that both $f(x) \to L$ and $g(x) \to M$ as $x \to p$. Then

1. $f(x) + g(x) \to L + M$ as $x \to p$,
2. $f(x) - g(x) \to L - M$ as $x \to p$,
3. $Kf(x) \to KL$ as $x \to p$,
4. $f(x)g(x) \to LM$ as $x \to p$,
5. $|f(x)| \to |L|$ as $x \to p$,
6. $\dfrac{f(x)}{g(x)} \to \dfrac{L}{M}$ as $x \to p$ provided that $M \neq 0$.

3.2.7 Theorem: limits of functions across an inequality   Suppose that f and g are two functions that possess limits as $x \to p$, and that $f(x) \leq g(x)$ for all values of x that are close to (but distinct from) p. Then

$$\lim_{x \to p} f(x) \leq \lim_{x \to p} g(x).$$

3.2.8 Theorem: a squeeze principle for function limits   Suppose that f, g and h are three functions, that both $f(x)$ and $h(x)$ possess *the same limit L* as $x \to p$, and that $f(x) \leq g(x) \leq h(x)$ for all values of x that are close to (but distinct from) p. Then also

$$g(x) \to L \text{ as } x \to p.$$

3.3 Continuity of real functions

Very informally, a function $f(x)$ is said to be continuous at a particular value of x – say, at $x = p$ – if its graph passes through the point $(p, f(p))$ without any kind of break. Furthermore, a function is said to be continuous if this holds good at every single point of its domain: that is, if the graph does not experience any kind of break at any point at which the function is defined.

Although, once again, that description is too vague to enable us to build up a proper mathematical investigation, it does connect with Section 3.2 in ways that strongly suggest how we can define the notion in a logically watertight way. Consider: in what ways can *f fail* to be continuous at $x = p$ by this informal criterion?

- $f(p)$ could fail to be defined at all, that is, p might not be in the domain of f:

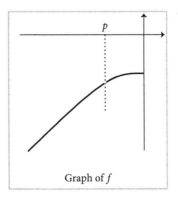

Graph of f

- $f(x)$ might fail to have a limit as $x \to p$:

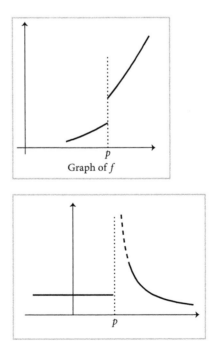

Graph of f

- Both $f(p)$ and $\lim_{x \to p} f(x)$ might exist, but be different numbers:

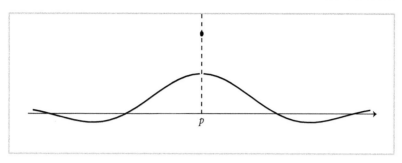

These being the scenarios that we need to avoid, a proper definition of continuity becomes almost inevitable:

3.3.1 Definition   A function $f : A \rightarrow B$ is *continuous at a point p of A if*[3]

$$\lim_{x \to p} f(x) = f(p)$$

or, in the alternative notation,

$$f(x) \rightarrow f(p) \text{ as } x \rightarrow p.$$

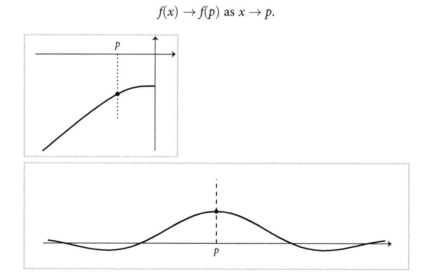

Since 3.2.1 and 3.2.3 offered two distinct ways to recognise function limits, they in turn create two distinct ways to define and work with continuity:

3.3.2 Alternative definition 1   A function $f : A \rightarrow B$ is *continuous at a point p* of A if, for every sequence $(x_n)_{n \in \mathbb{N}}$ in A such that $x_n \rightarrow p$, we have $f(x_n) \rightarrow f(p)$.

---

[3] This definition is *not quite universal*; if p is an isolated point of A, that is, p is the only element of A lying within some little open interval $(p - h, p + h)$ centred on p, then f is regarded as continuous at p even though its limit as $x \rightarrow p$ does not even make sense. This 'extreme' form of continuity does not arise in our later chapters, so we shall ignore its possibility.

3.3.3 **Alternative definition 2**   A function $f : A \to B$ is *continuous at a point p* of A if, for each $\varepsilon > 0$, we can find $\delta > 0$ such that

for every x in A such that $|x - p| < \delta$, we find that $|f(x) - f(p)| < \varepsilon$.

3.3.4 **Definition**   A function $f : A \to B$ is called *continuous* (or, if the emphasis is thought necessary, *continuous on A*) if it is continuous at every individual point of A.

3.3.5 **Notes**

- Common experience is that the condition in 3.3.2, often written more briefly in the form

$$x_n \to p \text{ (within } A\text{) implies } f(x_n) \to f(p)$$

and spoken as '*f preserves limits of sequences*', is usually the quickest and easiest way to handle continuity in practice.

- Amongst the many (and mostly rather predictable) results that can be established from any of these definitions of continuity, we shall need to use the three that we now list:

3.3.6 **Theorem: an algebra of continuous functions**

1. Adding, subtracting, scaling, multiplying and dividing continuous functions will always yield continuous functions *provided that* we avoid division by zero.

2. The composite of two (or more) continuous functions is always continuous.

3.3.7 **Theorem: continuity on bounded closed intervals**

1. A continuous function f on an interval of the form $[a, b]$ must itself be *bounded*: that is, there must be some constant K such that $|f(x)| < K$ for every x in $[a, b]$.

2. Indeed, any such function must actually attain a biggest value and a smallest value on that interval.

3.3.8 **The intermediate value theorem**   If J is a real interval (of any kind) and $f : J \to B$ is continuous, then its range $f(J)$ must also be an interval.

3.4 Differentiation of real functions

The gradient or slope of a straight line—say, the straight line through the points (a, b) and (c, d)—is given by $\dfrac{b - d}{a - c}$ (assuming $a \neq c$). The gradient of a curve, say, the graph of a function f at a typical point $(p, f(p))$ on that graph, could in principle be found by identifying the tangent (straight) line to the curve at that point and calculating its gradient; in practice, however, such a procedure is irremediably riddled with inaccuracy as well as fearfully time-consuming. The familiar alternative that differential calculus supplies is to determine instead the derivative (or derived function) f' of f, and to evaluate this at $x = p$. You will probably be aware of various almost mechanical algorithms for achieving this, and the present section intends a quick review of some of them together with a small selection of associated results.

3.4.1 **Definition**   Suppose that $f : A \rightarrow B$ is a real function and that p is an *interior* point of A (that is, p is the centre of some open interval that lies entirely within A). We say that f is *differentiable at p* if the limit

$$\lim_{x \to p} \frac{f(x) - f(p)}{x - p}$$

exists (or, equivalently, if the limit

$$\lim_{h \to 0} \frac{f(p + h) - f(p)}{h}$$

exists). This limit is called the *derivative* of f at p, usually denoted by $f'(p)$ and sometimes by $\dfrac{df}{dx}(p)$. It is easily (but informally) pictured as the slope or gradient of the graph of f at the point $(p, f(p))$ and of the tangent at that point, determined by approximating it using shorter and shorter secants of the graph.

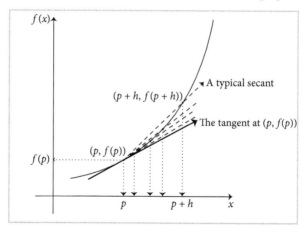

If (as is usually the case) $f'(p)$ exists at many (or even at all) points of A, so that f' becomes a function in its own right, it is often called the *derived function* of f.

The process of finding a derivative—however we choose to go about it—is called *differentiation*.

3.4.2 **Note**   It is easy to see that differentiable functions are always continuous: because if (in the above notation) $\dfrac{f(x) - f(p)}{x - p}$ possesses a limit L as $x \to p$, then also

$$f(x) = \left(\frac{f(x) - f(p)}{x - p}\right)(x - p) + f(p) \to (L)(0) + f(p) = f(p) \quad (\text{as } x \to p).$$

The converse is not true: even such a simple function as $f(x) = |x|$,

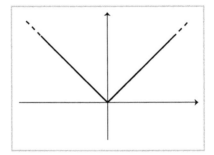

which is certainly continuous everywhere, is not differentiable at 0 because in this case the quantity $\dfrac{f(x) - f(0)}{x - 0}$ is 1 when x is positive but -1 when x is negative,

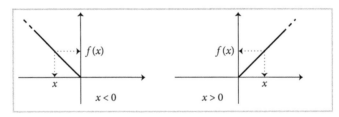

and clearly does not have a limit as $x \to 0$. Broadly speaking, continuous functions whose graphs feature sharp corners or cusps fail to be differentiable at any such points.

We recall a few standard derivatives:

3.4.3 **Proposition**   The derivatives of some basic functions, together with constraints on their validity where necessary, are set out in Table 3.1.

Table 3.1. Standard derivatives

$f(x)$	$f'(x)$	Condition?
x^a	ax^{a-1}	Any real constant a
$\sin x$	$\cos x$	
$\cos x$	$-\sin x$	
$\tan x$	$\sec^2 x$	x not of form $\frac{(2n+1)\pi}{2}, n \in \mathbb{Z}$
e^x	e^x	
$\log x$	$\dfrac{1}{x}$	$x > 0$
$\arcsin x$	$\dfrac{1}{\sqrt{1-x^2}}$	$-1 < x < 1$
$\arctan x$	$\dfrac{1}{1+x^2}$	

Here is a reminder of some basic calculus rules for differentiating combinations of functions:

3.4.4 Proposition   In each case, it is assumed that the derivatives on the right-hand side exist, either at appropriate points or on a set:

1. $(f(x) + g(x))' = f'(x) + g'(x)$
2. Where C is a constant, $(Cf(x))' = Cf'(x)$
3. $(f(x)g(x))' = f(x)g'(x) + g(x)f'(x)$
4. Provided that no division by zero is attempted, $\left(\dfrac{f(x)}{g(x)}\right)' = \dfrac{g(x)f'(x) - f(x)g'(x)}{(g(x))^2}$
5. Provided that the composite function makes sense, $(g(f(x)))' = g'(f(x))f'(x)$

Items 3, 4 and 5 here are generally called the *product rule*, the *quotient rule* and the *chain rule*. The chain rule is often written in the less modern alternative notation

$$\frac{dy}{dx} = \frac{dy}{du}\frac{du}{dx},$$

which certainly renders it easier to remember and, indeed, makes it appear to be mere common sense (although, in point of fact, it is by far the most difficult of the five rules for which to give a complete and comprehensive proof).

Derivatives have an important 'mean value' property in that, whenever a (continuous) function is differentiable throughout an interval, there must be a point where the derivative equals its average value across the entire interval:

3.4.5 **The 'first mean value theorem'**   If $f : [a, b] \rightarrow \mathbb{R}$ is continuous on its domain and differentiable on (a, b), then there must exist (at least) one point c in (a, b) for which

$$f'(c) = \frac{f(b) - f(a)}{b - a}.$$

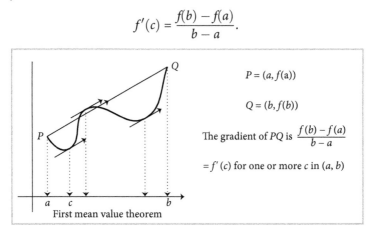

$P = (a, f(a))$

$Q = (b, f(b))$

The gradient of PQ is $\dfrac{f(b) - f(a)}{b - a}$

$= f'(c)$ for one or more c in (a, b)

First mean value theorem

As a small but important extension to paragraph 3.4.1, it is sometimes necessary to speak of a function being differentiable on a *closed* interval $[a, b]$ including at its endpoints. For a function $f : [a, b] \rightarrow \mathbb{R}$ that has such an interval as its domain, $f(x)$ makes no sense when $x < a$ nor when $x > b$ so, although 3.4.1's description of f' at the interior points is still fine, we can only investigate $f'(a)$ and $f'(b)$ by resorting to *one-sided limits*. In other words, we have the following:

3.4.6 **Definition**   If $f : [a, b] \rightarrow \mathbb{R}$, then we define

$$f'(a) = \lim_{x>a,\ x \to a} \frac{f(x) - f(a)}{x - a}$$

or, equivalently,

$$f'(a) = \lim_{h>0,\ h \to 0} \frac{f(a + h) - f(a)}{h}$$

provided that the limit exists. Again,

$$f'(b) = \lim_{x<b,\ x \to b} \frac{f(x) - f(b)}{x - b}$$

or, equivalently,

$$f'(b) = \lim_{h<0,\ h \to 0} \frac{f(b + h) - f(b)}{h}$$

provided that the limit exists. Then f is said to be *differentiable on* $[a, b]$ if its derivative exists at every point of $[a, b]$, including (in this sense) the endpoints a and b.

3.4.7 Note Differentiability on a closed interval implies continuity on that interval, for much the same reason as before. Notice, however, that it is perfectly possible for a function to be differentiable on $[a, b]$ and on an adjacent interval $[b, c]$ *and yet not be differentiable on their union* $[a, c]$: an easy illustration is provided by the function $f: [0, 2] \to \mathbb{R}$ specified by

$$f(x) = \begin{cases} x & \text{if } 0 \le x \le 1, \\ 2x - 1 & \text{if } 1 \le x \le 2. \end{cases}$$

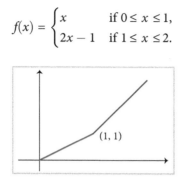

Noticing that the two formulae agree at $x = 1$, it is very routine to check that f is differentiable (with $f' = 1$ constantly) on $[0, 1]$, and that f is differentiable (with $f' = 2$ constantly) on $[1, 2]$. However, any attempt to calculate f' at the interior point 1 of $[0, 2]$ fails, because the *two-sided limit*

$$\lim_{x \to 1} \frac{f(x) - f(1)}{x - 1}$$

does not exist. Graphically speaking, this is witnessed by the (somewhat) sharp corner on the graph at the point $(1, 1)$.

3.4.8 Note When a function f has a derivative at every point of an interval, then the derivative f' is itself a function defined on that interval and we can sensibly ask whether it can be differentiated. If it can, then the derivative of f' is called the *second derivative* of f, usually written as f''. If f'' in its turn happens to be differentiable, then its derivative—the *third derivative* of f—is denoted by f''', and so on. However, the multiple dash notation gets very hard to read clearly at and beyond three, and cleaner notations are preferred. We shall opt for $f^{(3)}, f^{(4)}, f^{(5)}$ and so on to symbolise these *higher derivatives* of f whenever they are likely to be useful to us.

3.5 A very brief look at partial differentiation

When some quantity u is dependent upon (that is, is a function of) not *one* real variable but two or more, then there are a number of ways to consider differentiating it. Luckily for us, the type of differentiation involving multiple real variables that is most directly useful in basic complex analysis is the simplest possible of these: it amounts to pretending that all except one of the controlling variables are temporarily held constant, and then differentiating just as usual while only the one remaining variable is actually allowed to vary. To take a simple illustration: suppose that

$$u = u(x, y)$$

describes u as a function of two real variables x and y and, to make this initial discussion concrete, we shall take the particular formula

$$u(x, y) = e^{xy^3} \sin(x^2 + y^2).$$

- Consider that y is (temporarily) a constant, so that u now becomes a function of x only. Differentiating with respect to x (via the chain rule and the product rule) we obtain the derivative

$$e^{xy^3} \cos(x^2 + y^2)(2x) + e^{xy^3} (y^3) \sin(x^2 + y^2)$$

$$= e^{xy^3} (2x \cos(x^2 + y^2) + y^3 \sin(x^2 + y^2)).$$

- Secondly, consider this time that x is (temporarily) a constant, so that u now becomes a function of y only. Differentiating with respect to y, we obtain

$$e^{xy^3} \cos(x^2 + y^2)(2y) + e^{xy^3} (3xy^2) \sin(x^2 + y^2)$$

$$= ye^{xy^3} (2 \cos(x^2 + y^2) + 3xy \sin(x^2 + y^2)).$$

These one-at-a-time derivatives are called *partial derivatives*. All we need know of them at this stage is their proper definitions, the one-variable-at-a-time procedure illustrated above, and the standard symbols that denote them.

3.5.1 **Definition**   Suppose that $u = u(x, y)$ is a (real-valued) function of the two real variables x and y whose domain contains some disc centred on the point (p, q) in the coordinate plane. We define (provided that these limits do exist)

$$\frac{\partial u}{\partial x}(p, q) = \lim_{h \to 0} \frac{u(p + h, q) - u(p, q)}{h}$$

called the *partial derivative of u with respect to x at* (p, q), also denoted by $u_x(p, q)$, and

$$\frac{\partial u}{\partial y}(p, q) = \lim_{h \to 0} \frac{u(p, q + h) - u(p, q)}{h}$$

called the *partial derivative of u with respect to y at* (p, q), also denoted by $u_y(p, q)$.

When (as usually happens) they exist at every point throughout a region of the coordinate plane, then these partial derivatives are themselves functions of x and y whose domain is that region. It will then become valid to try partially differentiating these in turn, to seek higher-order partial derivatives (if and where they exist). Please take careful note of both of the standard notations:

$$\frac{\partial}{\partial x}\left(\frac{\partial u}{\partial x}\right) = \frac{\partial^2 u}{\partial x^2} = u_{xx},$$

$$\frac{\partial}{\partial y}\left(\frac{\partial u}{\partial x}\right) = \frac{\partial^2 u}{\partial y \partial x} = u_{xy},$$

$$\frac{\partial}{\partial x}\left(\frac{\partial u}{\partial y}\right) = \frac{\partial^2 u}{\partial x \partial y} = u_{yx} \text{ and}$$

$$\frac{\partial}{\partial y}\left(\frac{\partial u}{\partial y}\right) = \frac{\partial^2 u}{\partial y^2} = u_{yy}.$$

3.5.2 Example Find the two first-order and the four second-order partial derivatives of the function $u(x, y) = e^{2x} \sin(3y)$.

Solution

1.

$$\frac{\partial u}{\partial x} = 2e^{2x} \sin(3y) \quad \text{and} \quad \frac{\partial u}{\partial y} = 3e^{2x} \cos(3y).$$

2.

$$\frac{\partial^2 u}{\partial x^2} = \frac{\partial}{\partial x}\frac{\partial u}{\partial x} = 4e^{2x} \sin(3y),$$

$$\frac{\partial^2 u}{\partial y \partial x} = \frac{\partial}{\partial y}\frac{\partial u}{\partial x} = 6e^{2x} \cos(3y),$$

$$\frac{\partial^2 u}{\partial y^2} = \frac{\partial}{\partial y}\frac{\partial u}{\partial y} = -9e^{2x} \sin(3y) \quad \text{and}$$

$$\frac{\partial^2 u}{\partial x \partial y} = \frac{\partial}{\partial x}\frac{\partial u}{\partial y} = 6e^{2x} \cos(3y).$$

3.5.3 Example Find the four second-order partial derivatives of the function $u(x, y) = x^2 y \cos(x^3 + e^y)$.

Solution

1.

$$\frac{\partial u}{\partial x} = 2xy\cos(x^3 + e^y) - 3x^4 y\sin(x^3 + e^y) \quad \text{so}$$

2.

$$\frac{\partial^2 u}{\partial x^2} = 2y\cos(x^3 + e^y) - 6x^3 y\sin(x^3 + e^y) - 12x^3 y\sin(x^3 + e^y) - 9x^6 y\cos(x^3 + e^y)$$

$$= 2y\cos(x^3 + e^y) - 18x^3 y\sin(x^3 + e^y) - 9x^6 y\cos(x^3 + e^y) \quad \text{and}$$

3.

$$\frac{\partial^2 u}{\partial y\partial x} = 2x\cos(x^3 + e^y) - 2xye^y\sin(x^3 + e^y) - 3x^4\sin(x^3 + e^y) - 3x^4 ye^y\cos(x^3 + e^y)$$

$$= (2x - 3x^4 ye^y)\cos(x^3 + e^y) - (2xye^y + 3x^4)\sin(x^3 + e^y);$$

4.

$$\frac{\partial u}{\partial y} = x^2\cos(x^3 + e^y) - x^2 ye^y\sin(x^3 + e^y) \quad \text{so}$$

5.

$$\frac{\partial^2 u}{\partial y^2} = -x^2 e^y\sin(x^3 + e^y) - x^2(ye^y + e^y)\sin(x^3 + e^y) - x^2 ye^{2y}\cos(x^3 + e^y)$$

$$= -x^2 e^y(2 + y)\sin(x^3 + e^y) - x^2 ye^{2y}\cos(x^3 + e^y) \quad \text{and}$$

6.

$$\frac{\partial^2 u}{\partial x\partial y} = 2x\cos(x^3 + e^y) - 3x^4\sin(x^3 + e^y) - 2xye^y\sin(x^3 + e^y) - 3x^4 ye^y\cos(x^3 + e^y)$$

$$= (2x - 3x^4 ye^y)\cos(x^3 + e^y) - (3x^4 + 2xye^y)\sin(x^3 + e^y).$$

Calculations such as these can often become rather tedious, but the important points to note are

- all they require (apart from some patience) is a determination to treat the variables one at a time so that, at each given stage, all except one shall be viewed as constants and the differentiation process is carried out for one variable only;
- the observed fact, that $\frac{\partial^2 u}{\partial x\partial y}$ and $\frac{\partial^2 u}{\partial y\partial x}$ (for the functions in 3.5.2 and 3.5.3) turned out to be identical, happens almost always.

3.6 Exercises

(**Note:** Feel free to use the hyperbolic functions (see 2.5 Exercise 13) whenever they are useful.)

1. Find the two first-order and the four second-order partial derivatives of the function $u(x, y) = e^{ax} \cos(by)$ where a and b are (real) constants. Further,

   (i) check whether or not $\dfrac{\partial^2 u}{\partial x \partial y}$ and $\dfrac{\partial^2 u}{\partial y \partial x}$ are identical, and

   (ii) determine whether or not $\dfrac{\partial^2 u}{\partial x^2} + \dfrac{\partial^2 u}{\partial y^2}$ is zero under appropriate conditions on a and b if necessary.

2. Find the two first-order and the four second-order partial derivatives of the function $u(x, y) = \sin(ax) \sinh(by)$ where a and b are (real) constants. Further,

   (i) check whether or not $\dfrac{\partial^2 u}{\partial x \partial y}$ and $\dfrac{\partial^2 u}{\partial y \partial x}$ are identical, and

   (ii) determine whether or not $\dfrac{\partial^2 u}{\partial x^2} + \dfrac{\partial^2 u}{\partial y^2}$ is zero under appropriate conditions on a and b if necessary.

3. Find the four second-order partial derivatives of the function $u(x, y) = (x + y) \cos(xy)$ and determine whether or not $\dfrac{\partial^2 u}{\partial x \partial y}$ and $\dfrac{\partial^2 u}{\partial y \partial x}$ are identical.

4. Find the two first-order and the four second-order partial derivatives of each of the following functions $u = u(x, y)$, and decide whether or not

   • $\dfrac{\partial^2 u}{\partial x \partial y}$ and $\dfrac{\partial^2 u}{\partial y \partial x}$ are identical, and

   • $\dfrac{\partial^2 u}{\partial x^2} + \dfrac{\partial^2 u}{\partial y^2}$ is zero under appropriate conditions on the various constants a, b, c and d if necessary.

   (i) $ax^3 + bx^2y + cxy^2 + dy^3$,

   (ii) $x^3 + y^3 + ax^2 + bxy + cy^2$.

5. A real function $f = f(x, y)$ is said to be *harmonic* (on $\mathbb{R}^2$) if it satisfies the condition

$$\frac{\partial^2 f}{\partial x^2} + \frac{\partial^2 f}{\partial y^2} = 0$$

   for all real x and y. Given that the function $f(x, y) = ax^4 + bx^2y + cxy^2 + dy^4$ (where a, b, c and d are real constants) is harmonic throughout the plane, evaluate a, b, c and d.

6. Obtain the first-order partial derivatives of the functions $u = e^y(x \cos x + y \sin x)$ and $v = e^y(y \cos x - x \sin x)$. Take note of any coincidences that appear.

4 Complex functions

4.1 Introduction

Let's begin by agreeing that we know what is meant by subtracting one number from another, and by the difference between two numbers t and u (apart from a slight and rather unimportant ambiguity as to whether this means $t - u$ or $u - t$).

Now consider the statement 'the modulus of a number is its distance from zero, and the distance between two numbers is the modulus of their difference'.

There are two different standpoints from which we might read that statement. We could, on the one hand, start from an algebraic footing (knowing a formula for modulus) and perceive it as a definition of the geometric idea of distance. On the other hand, if we begin with our feet firmly planted in geometry (including an understanding of distance) we can view it as a way to define the algebraic notion of modulus. As usual, it doesn't actually matter which of the two attitudes you strike, for this is yet another small illustration of the way in which algebra and geometry cross-fertilise one another. There is no discord between them here: each provides additional or alternative insights into what the other wants to tell us.

This has implications for ideas such as limit and continuity, because we can conveniently express—for a start—the definition of limit of a sequence through the language of distance: a sequence $(t_n)_{n \in \mathbb{N}}$ converges to a limit L if, for each positive 'tolerance' ε, we can arrange that the distance between t_n and L shall be less than ε just by forcing n to reach or exceed some 'threshold' positive integer n_ε. Continuing, it is equally routine to re-express the idea of a function f having a limit L at a point p using the same distance-based language: namely, that for every $\varepsilon > 0$ we can force the distance between $f(t)$ and L to be less than ε just by making sure that t shall be closer to p in distance terms than some (carefully calculated) positive displacement δ (but nevertheless distinct from p). With that in place, continuity of f at p can continue to mean that the limit of $f(t)$ as t approaches p shall equal $f(p)$. Even differentiability of f at p as previously defined—that

$$\frac{f(t) - f(p)}{t - p}$$

shall possess a limit as t approaches p—presents no further difficulty provided that we can make sense of the division process displayed.

Notice that we haven't said whether the 'numbers' in this introductory discussion are real or complex. *This is because it doesn't matter*: once we have a clear idea

Integration with Complex Numbers. McCluskey and McMaster, Oxford University Press.
© Brian McMaster and Aisling McCluskey (2022). DOI: 10.1093/oso/9780192846075.003.0004

of distance between points—whatever kind of objects those points are—then the definitions of sequence limit, function limit and function continuity fall into place in the exact manner that we revised earlier, and so does that of differentiation provided also that the division procedure makes sense. This is the first of two key messages of the present chapter: that anyone who hoped/feared that the limits, continuity and differentiability of functions that operate with complex numbers would immediately present us with new and exciting challenges is going to be disappointed/relieved; the upcoming definitions for complex functions, and the initial theory that begins to build upon them, are virtually identical to those for real functions.[1]

4.2 Limits, continuity, differentiation (again)

4.2.1 Definitions   A *complex function* is a map from a subset A of $\mathbb{C}$ (the set of complex numbers) to a subset B of $\mathbb{C}$. That is, it is a procedure of some kind that allows us to determine, for each individual number z in A, a single 'corresponding' number in B. The sets A and B are called the *domain* of this function and its *codomain*. The group of symbols

$$f : A \to B$$

says that f is a complex function from domain A to codomain B (and then $f(z)$ denotes the element of B that corresponds to a typical element z of A). In most of the cases that concern us, the mapping will be made specific by stating some kind of formula telling us how to calculate $f(z)$ for each relevant z. When this is so, the default domain A (unless otherwise specified) will be the set of all complex z for which the formula makes sense (and delivers a complex number), and the default codomain will be the whole of $\mathbb{C}$.

Just as for real functions, there are two useful, distinct and completely interchangeable ways to define and to work with limits of complex functions:

4.2.2 Definition: complex function limits via sequences   We say that a complex function $f : A \to B$ *tends* (or *converges*) to a *limit* L in $\mathbb{C}$ as z tends (or

---

[1] Do keep in mind, however, that complex numbers do not support inequality ideas such as *less than, greater than, right, left, increasing, decreasing* which are very often useful for real numbers; so, for instance, once we are operating in the complex plane, we must avoid replacing conditions like $|t-p| < \delta$ by $p - \delta < t < p + \delta$ or $|f(t) - L| < \varepsilon$ by $L - \varepsilon < f(t) < L + \varepsilon$, or trying to break up $\lim_{z \to p} f(z)$ into separate considerations of 'left-hand' and 'right-hand' limits as z approaches p 'from the left' and 'from the right'. Any such attempt would simply not make sense for complex numbers and functions.

converges) to w if, for *every* sequence $(z_n)_{n\in\mathbb{N}}$ in $A \setminus \{w\}$ such that $z_n \to w$, we find that $f(z_n) \to L$. In this case we write

$$\lim_{z \to w} f(z) = L$$

or, equivalently,

$$f(z) \to L \text{ as } z \to w.$$

4.2.3 Example   Use 4.2.2 to determine the limit

$$\lim_{z \to 2} \frac{z^3 - 8}{z^5 - 32}.$$

Partial solution   The (default) domain is the whole of $\mathbb{C}$ except for the zeros of the bottom line, that is, the five distinct fifth roots of 32 (see paragraph 2.4.8):

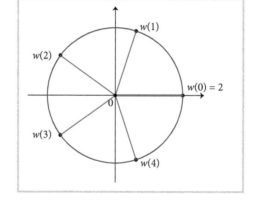

If (z_n) is any sequence in the domain such that $z_n \to 2$, then

$$f(z_n) = \frac{z_n^3 - 8}{z_n^5 - 32}$$

and, from that point on, an argument identical to that of paragraph 3.2.2 will show that $f(z) \to \frac{3}{20}$ as $z \to 2$ (and the fact that the arguments are identical is the most important point here).

4.2.4 Definition: complex function limits via 'epsilontics'   We say that a complex function $f : A \to B$ *tends* (or *converges*) to a *limit* L in $\mathbb{C}$ as z tends (or converges) to w if, for *every* positive real number ε we can find a positive number δ such that

for every z in A such that $0 < |z - w| < \delta$, we find that $|f(z) - L| < \varepsilon$.

In this case we (again) write $\lim_{z \to w} f(z) = L$ or, equivalently, $f(z) \to L$ as $z \to w$.

4.2.5 Example   With $f(z) = \dfrac{(z^2 + 1)^2}{|z - i|}$, use an epsilontic argument to verify that $f(z) \to 0$ as $z \to i$.

Solution   The domain of f is the whole of $\mathbb{C}$ except for i itself. Now

$$|f(z) - 0| = \left| \frac{(z^2 + 1)^2}{|z - i|} \right| = \left| \frac{(z + i)^2 (z - i)^2}{|z - i|} \right| = \frac{|z + i|^2 |z - i|^2}{|z - i|} = |z + i|^2 |z - i|.$$

If (as a first guess) we make sure that $|z - i| < 1$, then

$$|z + i| = |z - i + 2i| \le |z - i| + |2i| < 1 + 2 = 3, \text{ and therefore}$$

$$|z + i|^2 |z - i| < 3^2 |z - i| = 9|z - i|.$$

Given $\varepsilon > 0$, if we additionally make sure that $|z - i| < \varepsilon/9$, then this last $9|z - i|$ will be less than the given ε.

So if we define $\delta = \min\{1, \varepsilon/9\}$ (which is indeed greater than 0) then $0 < |z - i| < \delta$ implies that $|f(z) - 0| < \varepsilon$, as required.

4.2.6 Theorem: an algebra of limits for complex functions   Suppose that f and g are two complex functions, that K is a complex constant, and that both $f(z) \to L$ and $g(z) \to M$ as $z \to w$. Then

1. $f(z) + g(z) \to L + M$ as $z \to w$,
2. $f(z) - g(z) \to L - M$ as $z \to w$,
3. $Kf(z) \to KL$ as $z \to w$,
4. $f(z)g(z) \to LM$ as $z \to w$,
5. $|f(z)| \to |L|$ as $z \to w$,
6. $\dfrac{f(z)}{g(z)} \to \dfrac{L}{M}$ as $z \to w$ provided that $M \ne 0$.

Notice the absence of any theorem concerning how to take limits of complex functions across an inequality, or any sort of squeeze principle for complex function limits! Such ideas fail to make sense once the numbers underlying the discussion go complex.

It is occasionally useful to discuss limiting values of complex functions $f(z)$ *not* as z becomes very close to some centrally important value w, but, *on the contrary*, as z becomes very remote from such a central focus of attention. Limits such as these are called 'limits as z tends to infinity' or, equivalently, 'limits as $|z|$ tends to infinity'. Here is a way to express this idea precisely:

4.2.7 Definition   Suppose that f is a complex function whose domain A includes points arbitrarily far from zero, and that L is a complex number. Then we say that $f(z)$ *tends to L as* $|z|$ *tends to infinity* and write

$$f(z) \to L \text{ as } |z| \to \infty \quad \text{or} \quad \lim_{|z|\to\infty} f(z) = L$$

if, for each $\varepsilon > 0$, we can find some (large) positive number K_ε such that

$$|z| > K_\varepsilon \text{ implies } |f(z) - L| < \varepsilon.$$

(Informally, if we can force $f(z)$ as close to L as we can ever wish, just by taking z sufficiently far away from zero.)

4.2.8 **Example**   Show that the function

$$f(z) = e^{z-3|z|}$$

tends to 0 as $|z| \to \infty$.

Solution   Notice for a start that

$$\left|e^z\right| = \left|e^{\operatorname{Re}(z)+i\operatorname{Im}(z)}\right| = \left|e^{\operatorname{Re}(z)}\right|\left|e^{i\operatorname{Im}(z)}\right| = e^{\operatorname{Re}(z)}.$$

Notice secondly that $\operatorname{Re}(z - 3|z|) = \operatorname{Re}(z) - 3|z| \le |z| - 3|z| = -2|z|$ and that, in consequence,

$$\left|e^{z-3|z|}\right| = e^{\operatorname{Re}(z)-3|z|} \le e^{-2|z|}.$$

Given $\varepsilon > 0$, we can choose K_ε to be any positive number that exceeds $-\frac{1}{2}\log\varepsilon$. Then

$$|z| > K_\varepsilon \Rightarrow |z| > -\frac{1}{2}\log\varepsilon \Rightarrow -2|z| < \log\varepsilon \Rightarrow e^{-2|z|} < \varepsilon,$$

which in turn implies that $|f(z) - 0| < \varepsilon$. Hence the result.

4.2.9 **Definition**   A complex function $f : A \to B$ is *continuous at a point* $w \in A$ if [2]

$$\lim_{z\to w} f(z) = f(w)$$

or, in the alternative notation, if

$$f(z) \to f(w) \text{ as } z \to w.$$

From the two distinct ways of recognising complex function limits that we set out, there arise two useful distinct ways to define and work with continuity:

4.2.10 **Alternative definition 1**   A complex function $f : A \to B$ is *continuous at a point* $w \in A$ if, for every sequence $(z_n)_{n\in\mathbb{N}}$ in A such that $z_n \to w$, we have $f(z_n) \to f(w)$.

---

[2] More precisely, at a non-isolated point; see 3.2.5 (5) and the footnote in 3.3.1.

4.2.11 Alternative definition 2   A complex function $f : A \to B$ is *continuous at a point* $w \in A$ if, for each $\varepsilon > 0$, we can find $\delta > 0$ such that

for every z in A such that $|z - w| < \delta$, we find that $|f(z) - f(w)| < \varepsilon$.

4.2.12 Definition   A complex function $f : A \to B$ is called *continuous* (or, if the emphasis is thought necessary, *continuous on* A) if it is continuous at every individual point of A.

4.2.13 Notes

- Common experience is that the condition in 4.2.10, often written more briefly in the form

$$z_n \to w \text{ (within } A) \text{ implies } f(z_n) \to f(w)$$

and spoken as 'f *preserves limits of sequences*', is usually the quickest and easiest way to handle complex continuity in practice.

- Amongst the many (and mostly rather predictable) results that can be established from these definitions of complex continuity, we shall need to use the following three:

4.2.14 Theorem: an algebra of continuous complex functions

1. Adding, subtracting, scaling, multiplying and dividing continuous complex functions will always yield continuous complex functions *provided that* we avoid division by zero.

2. The composite of two (or more) continuous complex functions is always continuous.

3. Of course, constant functions are (trivially) continuous, and so is the so-called identity function $f(z) = z$. It follows from this very modest starting point, using item 1 above, that all (complex) polynomials are everywhere continuous, and so are all (complex) rational functions, that is, all functions of the form $\dfrac{p_1(z)}{p_2(z)}$ where p_1 and p_2 are complex polynomials *except, of course,* at the points where p_2 takes the value 0 (since the rational function is not even defined at such points).

Recalling that a continuous *real* function on a closed bounded interval will always be bounded, we next wish to obtain an analogous result for complex functions. The part of this investigation that requires most preparation is clarifying precisely what is meant by calling a subset of $\mathbb{C}$ *closed*.

4.2.15 Notation   For $z_0 \in \mathbb{C}$ and $r > 0$:

- the notation $C(z_0, r)$ denotes the circle of centre z_0 and radius r, more precisely described as $\{z \in \mathbb{C} : |z - z_0| = r\}$, and

- the notation $D(z_0, r)$ denotes the region lying inside that circle, the *open disc* of centre z_0 and radius r, that is, the set $\{z \in \mathbb{C} : |z - z_0| < r\}$.

- We also occasionally make use of the corresponding *closed disc* $\overline{D}(z_0, r)$ formed by combining the two, that is, the set $\{z \in \mathbb{C} : |z - z_0| \leq r\}$.

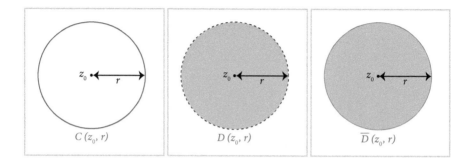

$C(z_0, r)$ $D(z_0, r)$ $\overline{D}(z_0, r)$

4.2.16 Definition A non-empty subset G of $\mathbb{C}$ is called *open* (or *an open set*) if every point of G is the centre of some open disc that is small enough to lie entirely inside G.[3]

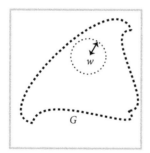

Informally speaking, an open set is one that does not include any of its boundary points. For instance, each open disc $D(z_0, r)$ is an open set because, for any point w belonging to it, the (smaller) open disc $D(w, r - |w - z_0|)$ is totally contained in $D(z_0, r)$. In contrast, $\overline{D}(z_0, r)$ is not an open set: if w is a point lying on its boundary circle $C(z_0, r)$, then any disc centred on w, however small we make its radius, will include points *outside* the bounding circle and therefore not in $\overline{D}(z_0, r)$.

---

[3] To express that formally: if for each $w \in G$ there exists some radius $r_w > 0$ such that $D(w, r_w) \subseteq G$.

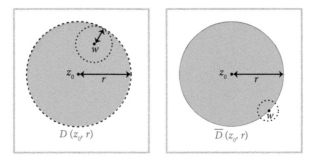

$D(z_0, r)$ $\overline{D}(z_0, r)$

4.2.17 Definition   The complement in $\mathbb{C}$ of an open set (in $\mathbb{C}$) is called a *closed set*. For instance, $\overline{D}(a, r)$ is closed for any choice of a and r.

Closed sets have a property that is valuable whenever you need to take limits of sequences: because if K is closed and $(z_n)_{n \in \mathbb{N}}$ is any convergent sequence of points in K, then the limit of (z_n) must also belong to K. Indeed, as we shall now show, this condition is *characteristic* of closed sets. (The next three proofs are quite different in style from most of the others in this text, and you might want to leave them out on a first reading, but take care to understand the results.)

4.2.18 Proposition   Suppose that K is a subset of $\mathbb{C}$. Then K is closed if and only if:

whenever $(z_n)_{n \in \mathbb{N}}$ is a convergent sequence of elements of K, then $\lim z_n \in K$.

Proof

1. Suppose K is closed. Consider a typical convergent sequence (z_n) of elements of K and let w denote its limit.
   If w does not belong to K then w belongs instead to $\mathbb{C} \setminus K$ which is an open set. Its openness tells us that there is an open ball $D(w, r)$ centred on w and so small that it lies entirely inside $\mathbb{C} \setminus K$. In particular, $D(w, r)$ cannot contain any of the points z_n.
   Yet this *contradicts $z_n \to w$*, since $|z_n - w|$ tends to zero and must therefore be eventually smaller than r.

2. Suppose K is not closed, that is, $\mathbb{C} \setminus K$ is not open.
   Then there must be a point w in $\mathbb{C} \setminus K$ such that, no matter how small we choose its radius r, the ball $D(w, r)$ cannot fit entirely inside $\mathbb{C} \setminus K$.
   In particular, for each positive integer n the open ball $D(w, \frac{1}{n})$ is not contained in $\mathbb{C} \setminus K$, and therefore we can choose a point z_n that is in $D(w, \frac{1}{n})$ but not in $\mathbb{C} \setminus K$.
   So every z_n belongs to K, and $|z_n - w| < \frac{1}{n}$ which tends to zero.

We have thus constructed a convergent sequence of elements of K whose limit does not belong to K, and the condition in the Proposition fails.

For just one theorem[4] in Chapter 7 we shall need to know how 'nests' of closed sets behave:

4.2.19 Lemma   Let $K_1 \supseteq K_2 \supseteq K_3 \supseteq K_4 \supseteq K_5 \supseteq \cdots$ be a sequence of non-empty closed sets, each one of which contains the next one, and suppose that K_1 is bounded (that is, it lies entirely within some finite distance from zero). Then there is a point z that is common to *every* set K_m in the sequence (in other words, the intersection of all of the sets K_m is not empty).

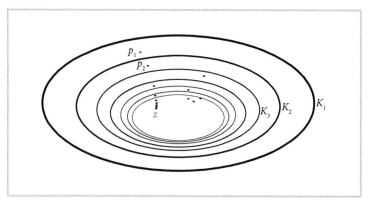

Proof   Choose, from each of the sets K_n, a point p_n in any way you please. The sequence $(p_n)_{n\in\mathbb{N}}$ which this accidentally creates is bounded (because K_1 is), so Bolzano-Weierstrass tells us that one of its subsequences converges (to some limit z). For each positive integer m, the set K_m includes all but the first few terms of this subsequence (because of the way these sets are nested inside one another) and therefore, since it is a closed set and early terms have no effect on a sequence's limit, it includes their limit. That is, $z \in K_m$ for every $m \in \mathbb{N}$.   ■

Recall that a complex sequence (z_n) is said to be bounded if there is some positive constant M such that $|z_n| < M$ for every term in the sequence. It is easy to adjust that definition to apply also to sets of complex numbers (as, indeed, we just did in 4.2.19) and also to complex functions.

4.2.20 Definition

1. A set S of complex numbers is called *bounded* or *a bounded set* if there is some positive constant M such that $|z| < M$ for every element z of S.

---

[4] Paragraph 7.3.1.

2. A complex function $f : A \rightarrow \mathbb{C}$ is called *bounded* or *a bounded function* if there is some positive constant M such that $|f(z)| < M$ for every z in A (equivalently, if its range $f(A)$ is a bounded set).

4.2.21 Theorem: continuity on closed bounded sets

1. A continuous complex function on a closed bounded subset of $\mathbb{C}$ must be bounded.

2. Indeed, *the modulus of* any such function must actually attain a biggest value and a smallest value.

Proof

1. Suppose that $f : A \rightarrow \mathbb{C}$ is continuous and that A is a closed bounded subset of $\mathbb{C}$. If f *were not bounded* then, in particular, for each positive integer n, the condition '$|f(z)| < n$' could not be true for all of the points z of A, so there must exist z_n in A for which $|f(z_n)| \geq n$. Since A is bounded, Bolzano-Weierstrass tells us that the bounded sequence (z_n) has a subsequence (z_{n_k}) that converges to some limit w. Since A is closed, 4.2.18 tells us that w belongs to A. Since f is continuous everywhere in A, $f(z_{n_k}) \rightarrow f(w)$, so $(f(z_{n_k}))$ is a convergent sequence *and hence bounded*. Yet $|f(z_{n_k})| \geq n_k \geq k$ where we can let k tend to infinity, *contradicting* the boundedness of $(f(z_{n_k}))$.

2. Continuing from part 1, let M_0 denote the supremum of all the (real) numbers $|f(z)|$ for $z \in A$. We have $|f(z)| \leq M_0$ for every $z \in A$ but, for each positive integer n, there is a point z_n in A for which $M_0 - \frac{1}{n} < |f(z_n)|$. As before, the bounded sequence (z_n) has a convergent subsequence (z_{n_k}) whose limit w belongs to A. Continuity of f yields $f(z_{n_k}) \rightarrow f(w)$, and therefore $|f(z_{n_k})| \rightarrow |f(w)|$. Yet it is clear from $M_0 - \frac{1}{n_k} < |f(z_{n_k})| \leq M_0$ that the limit of $|f(z_{n_k})|$ is actually M_0, that is, $|f(w)| = M_0$. So M_0 is not just the supremum of the values of $|f|$ but the greatest value among them.

   Similarly, the infimum of the values of $|f|$ is the least value among them.

4.2.22 Definition

Suppose that $f : A \rightarrow \mathbb{C}$ is a complex function and that w is an *interior* point[5] of A (that is, w is the centre of some open disc that lies entirely within A). We say that f is *differentiable at w* if the limit

$$\lim_{z \to w} \frac{f(z) - f(w)}{z - w}$$

exists (or, equivalently, if the limit

$$\lim_{h \to 0} \frac{f(w + h) - f(w)}{h}$$

---

[5] Compare this with the very similar definition we gave in 3.4.1 of an interior point of a subset of $\mathbb{R}$.

exists). This limit is called the *derivative* of *f* at *w*, usually denoted by $f'(w)$ and sometimes by $\frac{df}{dz}(w)$. If (as is usually the case) $f'(w)$ exists at many (or even at all) points of *A*, so that f' becomes a complex function in its own right, it is often called the *derived function* of *f*. The process of finding a derivative—however we choose to go about it—is called *differentiation*.

4.2.23 **Notes**

- It is easy to see, just as in the real case, that differentiable functions are always continuous. The converse is not true: it is a useful exercise to verify that the continuous complex function $f(z) = |z|$ is not differentiable at 0. (Indeed, rather more surprisingly, it is not differentiable anywhere in $\mathbb{C}$.)

- The product rule, quotient rule and chain rule for differentiating more complicated real functions are valid for complex functions also, and for the same reasons.

- Therefore we can build up a catalogue of basic differentiable complex functions such as constants, z, z^2, polynomials and rational functions (except of course at points where they fail to be defined due to the bottom line going zero). Likewise, functions such as e^z, sin z and cos z, defined as sums of power series that we know can be differentiated term by term, are differentiable and with known, familiar derivatives.

- Higher derivatives of complex functions, when they exist, are defined and annotated just as in the real case.

4.3 **Cauchy–Riemann**

When $f : A \to \mathbb{C}$ is a complex function, each of its values $f(z)$ is a complex number and therefore possesses a real part and an imaginary part. If (as is common practice) we denote these by *u* and *v*, so that $f(z) = u + iv$, then *u* and *v* will vary with *z* (that is, be functions of *z*) and so it would be perfectly correct to write $f(z) = u(z) + iv(z)$. Yet *z* in its turn has real and imaginary parts (let us call them *x* and *y* as usual) and so both *u* and *v* will vary with both *x* and *y*. The (rather clumpy-looking) notation

$$f(z) = u(x, y) + iv(x, y)$$

tries to make all these points at once: on the left-hand side we are thinking exclusively in terms of complex numbers, in which the complex quantity $f(z)$ is thought of as depending on the variable complex number *z*, while on the right-hand side we are seeking to 'get real', breaking everything down into the real quantities *u* and *v*, each of which is controlled by the two real numbers *x* and *y* for which $z = x + iy$.

This would be a dry and pointless exercise in slightly ambiguous notation, were it not for the fact that something interesting and important is about to emerge from it. Each of the real two-variable functions u and v may, of course, have partial derivatives with respect to both x and y (recall Section 3.5) which, as was remarked in paragraph 3.5.1, we commonly denote either by u_x, u_y, v_x and v_y or, if preferred, by

$$\frac{\partial u}{\partial x}, \frac{\partial u}{\partial y}, \frac{\partial v}{\partial x} \text{ and } \frac{\partial v}{\partial y}.$$

We set out to show that differentiability of the complex function f is intimately bound up with certain 'coincidences' between these four partial derivatives—coincidences known as the *Cauchy–Riemann equations*.

4.3.1 Theorem   Suppose that the complex function f is differentiable at the point $w = a + ib$ in the complex plane, and that we express it in terms of its real and imaginary parts as

$$f(z) = u(x, y) + iv(x, y).$$

Then, at the point (a, b), the four partial derivatives exist and

$$\frac{\partial u}{\partial x} = \frac{\partial v}{\partial y} \text{ and } \frac{\partial v}{\partial x} = -\frac{\partial u}{\partial y}.$$

Proof   We shall begin by confirming the *existence* of some partial derivatives at the point $w = (a, b) = a + ib$. The derivative $f'(w)$ is the limit of

$$\frac{f(z) - f(w)}{z - w} \quad (\text{as } z \to w)$$

so, in particular, for every non-zero *real* number h we have

$$\frac{f(w + h) - f(w)}{(w + h) - w} = \frac{u(a + h, b) + iv(a + h, b) - (u(a, b) + iv(a, b))}{h}$$

$$= \frac{u(a + h, b) - u(a, b)}{h} + i\frac{v(a + h, b) - v(a, b)}{h}$$

converging to $f'(w)$ as $h \to 0$. Yet the limit of the last display is, by definition, $\frac{\partial u}{\partial x} + i\frac{\partial v}{\partial x}$ evaluated at (a, b) so, at least, these two partial derivatives exist.

(Having obtained part of what we wanted by approaching w 'horizontally', the strategy now is to rerun the argument with a 'vertical' approach and then face up to the fact that the limits have to be the same in both cases.)

The derivative $f'(w)$ is the limit of

$$\frac{f(z) - f(w)}{z - w} \quad (\text{as } z \to w)$$

so, in particular, for every non-zero *real* number h we have

$$\frac{f(w + ih) - f(w)}{(w + ih) - w} = \frac{u(a, b + h) + iv(a, b + h) - (u(a, b) + iv(a, b))}{ih}$$

$$= \frac{u(a, b + h) - u(a, b)}{ih} + i\frac{v(a, b + h) - v(a, b)}{ih}$$

$$= -i\frac{u(a, b + h) - u(a, b)}{h} + \frac{v(a, b + h) - v(a, b)}{h}$$

converging to $f'(w)$ as $h \to 0$. Yet the limit of the last display is, by definition, $-i\frac{\partial u}{\partial y} + \frac{\partial v}{\partial y}$ evaluated at (a, b) so these two partial derivatives exist also.

In addition, $f'(w) = \frac{\partial u}{\partial x} + i\frac{\partial v}{\partial x} = -i\frac{\partial u}{\partial y} + \frac{\partial v}{\partial y}$ all evaluated at (a, b).

Equating real and imaginary parts, we get $\frac{\partial u}{\partial x} = \frac{\partial v}{\partial y}$ and $\frac{\partial v}{\partial x} = -\frac{\partial u}{\partial y}$. ∎

4.3.2 Notes

- It is worth stressing that this proof gives us a formula for the derivative of a (differentiable) complex function in terms of its real and imaginary parts, namely

$$f'(w) = \frac{\partial u}{\partial x} + i\frac{\partial v}{\partial x}$$

in the above notation (as well as an alternative one in which the partial derivatives are with respect to y instead of x).

- The Cauchy–Riemann equations quite often give us an easy way to show that a complex function is *not* differentiable. For instance, if we assume for a moment that the function $f(z) = |z|^2$ were differentiable at a non-zero point w then, writing $f(z) = u + iv = x^2 + y^2$ under the same notational conventions that we employed in Theorem 4.3.1 and its proof, we see that $u = x^2 + y^2$ and that $v = 0$ identically, so $\frac{\partial u}{\partial x} = 2x$, $\frac{\partial u}{\partial y} = 2y$, $\frac{\partial v}{\partial x} = 0$ and $\frac{\partial v}{\partial y} = 0$ (identically). The Cauchy–Riemann equations embodied in 4.3.1 now give $2x = 0$ and $2y = 0$, contradicting our assumption that w is non-zero. In other terms, $|z|^2$ cannot be differentiable at any point in $\mathbb{C}$ except, perhaps, at 0.

- The converse of 4.3.1 is *not quite true*. That is, $\dfrac{\partial u}{\partial x} = \dfrac{\partial v}{\partial y}$ and $\dfrac{\partial v}{\partial x} = -\dfrac{\partial u}{\partial y}$ (at a point (a, b)) are not on their own sufficient to guarantee that $f(z) = u + iv$ shall be differentiable at $a + ib$. A fairly easy way to see this is to check via the definition that the complex function $f(z) = \dfrac{z^5}{|z|^4}$ (supplemented by $f(0) = 0$) is not differentiable at zero, and yet it does satisfy the Cauchy–Riemann equations at that point (see Exercise 18 at the end of this chapter).

- This failure of 4.3.1 to have a simple-minded converse is less of a practical difficulty than it might appear to be. Much of the reason for this is just the way in which differentiability is defined: for remember that, in order to discuss whether a function f is differentiable at a point w, we first of all need $f(z)$ to be defined (at least) everywhere in some open disc centred on w. It follows that differentiability is normally most interesting and most useful *not* at a single point, but throughout an open disc or a union of open discs—in other words, throughout an open set. And it turns out to be *very nearly true* that if the Cauchy–Riemann equations hold at every point in an open set, then the complex function will be differentiable everywhere in that open set. (The missing link in that 'very nearly' is an additional condition that the partial derivatives should also be continuous—a circumstance that is usually easy to confirm. So a useful if-and-only-if variant of 4.3.1 runs as follows:)

4.3.3 Theorem   Suppose that G is an open subset of the domain of a complex function f, and that (in the notation where $f(z) = u + iv$, u and v being real) the four partial derivatives

$$\frac{\partial u}{\partial x}, \frac{\partial u}{\partial y}, \frac{\partial v}{\partial x}, \frac{\partial v}{\partial y}$$

exist and are continuous throughout G. Then f is differentiable on G if and only if the Cauchy–Riemann equations hold at every point of G.

Proof   Suppose the Cauchy–Riemann equations hold throughout G. Take a typical point $z_0 = a + ib = (a, b)$ in G and a small disc $D(z_0, r)$ contained in G, and consider another point $z = (a + h, b + k)$ in that disc. Our task is to show that $(f(z) - f(z_0))/(z - z_0)$ converges to a limit (namely $f'(z_0)$) as $z \to z_0$, in other words, that we can express

$$\frac{f(z) - f(z_0)}{z - z_0} = f'(z_0) + \varepsilon$$

for some error term ε that tends to zero as the radius r of our little disc shrinks to zero. We begin by working separately on the real part $u(x, y)$ of $f(z)$.

First consider moving 'horizontally' from (a, b) to $(a + h, b)$ in the complex plane. Only the x-coordinate is changing, so any mention of *derivative* will mean the partial derivative with respect to x. Invoking the first mean value theorem (paragraph 3.4.5) we find that $u(a + h, b) = u(a, b) + h\dfrac{\partial u}{\partial x}(p)$ where p is a point somewhere on the line from (a, b) to $(a + h, b)$. The continuity of $\dfrac{\partial u}{\partial x}$ tells us that $\dfrac{\partial u}{\partial x}(p) = \dfrac{\partial u}{\partial x}(a, b) + \varepsilon_1$ where the error term here, ε_1, tends to zero as the disc shrinks. So, at this stage, we have

$$u(a + h, b) = u(a, b) + h\left(\frac{\partial u}{\partial x}(a, b) + \varepsilon_1\right).$$

Next, imagine repeating the argument of the previous paragraph but moving 'vertically' from $(a+h, b)$ to $(a+h, b+k)$. This time it is the y-coordinate that alters, so the relevant derivative will be $\dfrac{\partial u}{\partial y}$ but nothing else changes. The conclusion will therefore be

$$u(a + h, b + k) = u(a + h, b) + k\left(\frac{\partial u}{\partial y}(a, b) + \varepsilon_2\right)$$

for another error term ε_2 that tends to zero as r does.

Combining the two displays gives us

$$u(a + h, b + k) = u(a, b) + h\left(\frac{\partial u}{\partial x}(a, b) + \varepsilon_1\right) + k\left(\frac{\partial u}{\partial y}(a, b) + \varepsilon_2\right).$$

Using exactly the same argument on v in place of u, we obtain

$$v(a + h, b + k) = v(a, b) + h\left(\frac{\partial v}{\partial x}(a, b) + \varepsilon_3\right) + k\left(\frac{\partial v}{\partial y}(a, b) + \varepsilon_4\right)$$

where $\varepsilon_3 \to 0$ and $\varepsilon_4 \to 0$, and all the partial derivatives are evaluated at $(a, b) = z_0$.

Putting u and v together since $f = u + iv$, it now follows that

$$f(z) - f(z_0) = h\left(\frac{\partial u}{\partial x} + \varepsilon_1\right) + k\left(\frac{\partial u}{\partial y} + \varepsilon_2\right) + i\left[h\left(\frac{\partial v}{\partial x} + \varepsilon_3\right) + k\left(\frac{\partial v}{\partial y} + \varepsilon_4\right)\right]$$

(*and at last we get to use the Cauchy–Riemann equations:*)

$$= h\left(\frac{\partial u}{\partial x} + \varepsilon_1\right) + k\left(-\frac{\partial v}{\partial x} + \varepsilon_2\right) + i\left[h\left(\frac{\partial v}{\partial x} + \varepsilon_3\right) + k\left(\frac{\partial u}{\partial x} + \varepsilon_4\right)\right]$$

which, as the disc shrinks, can be made as close as we please to

$$h\frac{\partial u}{\partial x} - k\frac{\partial v}{\partial x} + ih\frac{\partial v}{\partial x} + ik\frac{\partial u}{\partial x}$$

$$= (h + ik)\frac{\partial u}{\partial x} + (h + ik)i\frac{\partial v}{\partial x}$$

$$= (h + ik)\left(\frac{\partial u}{\partial x} + i\frac{\partial v}{\partial x}\right) = (z - z_0)\left(\frac{\partial u}{\partial x} + i\frac{\partial v}{\partial x}\right).$$

In other words, $\dfrac{f(z) - f(z_0)}{z - z_0}$ does possess a limit (as the disc enclosing z and z_0 shrinks down to its centre) and the limit (the derivative of f) is $\dfrac{\partial u}{\partial x} + i\dfrac{\partial v}{\partial x}$ (or, equivalently via Cauchy–Riemann, $\dfrac{\partial v}{\partial y} - i\dfrac{\partial u}{\partial y}$ if preferred).

The converse implication is immediate from 4.3.1. ∎

4.3.4 Note and definition   This is the first result we have met so far in which *differentiability* of a function and *openness* of a set in its domain have come together in a powerful way, and we shall encounter many more. Complex analysis is particularly interested in such functions, and it has several names[6] for them; we shall use the term *regular*. That is, if $f : A \to \mathbb{C}$ and G is an open subset of A and f is differentiable at every point of G, we say that f is *regular on G*.

4.4 Surprises!

Up to this point, Chapter 4 has been stressing that analytic ideas in the study of complex functions either are more-or-less identical to the corresponding concepts in real analysis or else can be reduced (as seen, for example, in the Cauchy–Riemann section) to such material. If this were universally true, of course, there would be no point in developing complex analysis. So it is high time that we flagged up the second key theme of the chapter: that, just sometimes, the behaviour of calculus notions that we thought we had thoroughly understood in the real case becomes startlingly different once we allow the underlying number structure to become complex. Proofs in this section will mostly be postponed to a later chapter in which we develop appropriate methods; the emphasis here is merely on what real analysis leads us to expect, and how complex analysis can utterly confound that expectation.

4.4.1 Note   As we have pointed out, it is easy to find a real function that is everywhere continuous, but not everywhere differentiable. The modulus function, for instance, is continuous on $\mathbb{R}$ but fails to have a derivative at 0. So if we define another function $f_1 : \mathbb{R} \to \mathbb{R}$ by the formula

---

[6] Including *analytic* and *holomorphic*.

$$f_1(x) = \begin{cases} -\frac{1}{2}x^2 & \text{if } x \le 0 \\ \frac{1}{2}x^2 & \text{if } x > 0 \end{cases}$$

then (provided we take a little extra care about what happens at $x = 0$) it is easy to check that $f_1'(x) = |x|$ throughout the real line. Thus we have found a simple real function that is differentiable everywhere once, but not twice. Then consider $f_2 : \mathbb{R} \to \mathbb{R}$ given by

$$f_2(x) = \begin{cases} -\frac{1}{6}x^3 & \text{if } x \le 0 \\ \frac{1}{6}x^3 & \text{if } x > 0. \end{cases}$$

With again a quantum of caution at 0, we can see that f_2 is differentiable everywhere, and that its derivative is exactly f_1; so f_2 is differentiable twice, but not thrice. The next step

$$f_3(x) = \begin{cases} -\frac{1}{24}x^4 & \text{if } x \le 0 \\ \frac{1}{24}x^4 & \text{if } x > 0 \end{cases}$$

devises a function that is differentiable three times but not four times. Clearly there is nothing to stop us from continuing inductively: *for every positive integer n there are real functions that are differentiable n times everywhere on the real line, but not n + 1 times.*

A reasonable person might expect the greater complexity of the complex number system to allow even wilder behaviour in such an arena, but the truth is precisely the opposite:

4.4.2 Theorem   If a complex function can be differentiated once at every point of an open set G, then it can be differentiated as often as you wish at every point of G.

Proof   See Chapter 8 (paragraphs 8.2.3 and 8.2.2).

4.4.3 Note   One of the essential and powerful applications of real calculus is Taylor's theorem: that, *in most cases,* a 'complicated' or subtle function such as $\cos x$ or e^x or $(1 + x)^{\frac{3}{4}}$ can be perfectly represented[7] by the sum of a power series that generally offers much easier ways to deal with it. You may well have encountered, for example, 'Taylor expansions' such as

$$\sin x = x - \frac{1}{3!}x^3 + \frac{1}{5!}x^5 - \frac{1}{7!}x^7 + \cdots,$$

$$\exp(x) = e^x = 1 + x + \frac{1}{2!}x^2 + \frac{1}{3!}x^3 + \frac{1}{4!}x^4 + \frac{1}{5!}x^5 + \cdots,$$

[7] We flagged up a few instances of representations like this in 1.4.9.

$$\arctan x = x - \frac{1}{3}x^3 + \frac{1}{5}x^5 - \frac{1}{7}x^7 + \cdots .$$

Unfortunately, however, the words *in most cases* are necessary here, because there are perfectly reasonable real functions for which the Taylor approximation process[8] does not work. An example is the function $f : \mathbb{R} \rightarrow \mathbb{R}$ defined by the formula

$$f(x) = \begin{cases} \exp\left(-\dfrac{1}{x^2}\right) & \text{if } x \neq 0, \\ 0 & \text{if } x = 0. \end{cases}$$

It takes quite a lot of effort and patience to work out all the higher derivatives of this function and to compile its Taylor series, but anyone who succeeds in this enterprise is doomed to disappointment, because the Taylor series turns out to be

$$0 + 0 + 0 + 0 + 0 + 0 + \cdots$$

and, as you can see, it completely fails even to try to describe how the function $f(x)$ itself varies with the values of x!

Once again, a reasonable (and slightly pessimistic) observer could anticipate that complex functions might manifest even more dramatic failures of what is essentially the same approximation process; and once again, that pessimism turns out to be misplaced:

4.4.4 Theorem   For any complex function f that is defined and differentiable on an open disc, the Taylor series for f (calculated in exactly the same way as is done for real functions) converges to $f(z)$ at every point z of that disc.

Proof   See Chapter 8, paragraph 8.2.3.

4.4.5 Note   Lastly for the present, think what you can confidently predict about a real function once you know that it is differentiable everywhere in $\mathbb{R}$ and also bounded there. The short answer is: *almost nothing*. There is such a dazzling diversity of bounded and universally differentiable real functions—consider just for a start $a \sin(bx + c)$ for all possible values of a, b and c, $\dfrac{1}{1 + x^2}$ and variations on that, $\arctan x$, all the possible combinations of these functions—that it is difficult to imagine any sensible statement one could make that had to be true for all of them. The proposal is almost certainly unrealisable.

And presumably, everywhere differentiable bounded *complex* functions will show an even greater variety of behaviour, an even more outrageous range of diversity. . . right?

Wrong:

---

[8] We shall look more carefully at the process itself in Chapter 8; the point being made here is that it works more smoothly in $\mathbb{C}$ than it does in $\mathbb{R}$.

4.4.6 Liouville's theorem   The only bounded complex functions that are every-where differentiable are the constant functions.

Proof   See Chapter 8, paragraph 8.2.5.

Even for the experienced mathematician there is a trace of mystery surrounding the fact that, symbol for symbol, the definition of differentiability for real functions and the definition of differentiability for complex functions appear to be identical and yet, somehow, it tells you a lot more about complex functions than it does about real functions. The great benefit of this mystery is that, as we shall show, *complex analysis* in many of its aspects (and despite the everyday meaning and tone of those words) is simpler, cleaner, more elegant than its real co-discipline. Paragraphs 4.4.2, 4.4.4 and 4.4.6 can serve as indicators of *the surprising simplicity of complexity*.

4.5 Exercises

[Unless otherwise indicated, u and v denote the real and imaginary parts of the relevant *function*, whereas x and y are those of the generic *complex number z*. Feel free to use the hyperbolic functions of 2.5 Exercise 13 whenever they are useful.]

1. Identify the real and imaginary parts of $\sin z$ and of $\cos z$ (where $z = x + iy$ is a typical complex number).

2. Show that there is a complex number whose cosine is 2; show that there is a complex number whose sine is $4i$.

3. Find *all* solutions in $\mathbb{C}$ of $\sin z = 0$ and *all* solutions in $\mathbb{C}$ of $\cos z = 0$.

4. Show that $\bar{z}$ is not differentiable anywhere in $\mathbb{C}$. More generally, determine the condition on the real constants a and b that ensures that the real and imaginary parts of the complex function f defined by the formula

$$f(z) = a \operatorname{Re}(z) + ib \operatorname{Im}(z), \quad z \in \mathbb{C},$$

satisfy the Cauchy–Riemann equations.

5. Determine whether or not the real and imaginary parts of the complex function f defined by the formula

$$f(z) = \overline{ze^{i\bar{z}}}, \quad z \in \mathbb{C}$$

satisfy the Cauchy–Riemann equations on $\mathbb{C}$.

6. Do the real and imaginary parts of $\cos(i\bar{z})$, $z \in \mathbb{C}$, satisfy the Cauchy–Riemann equations on an open subset of $\mathbb{C}$?

7. Given two real constants a and b, identify a complex function that is differentiable everywhere in $\mathbb{C}$ and whose imaginary part at $z = x + iy$ is $a(x^2 - y^2) + b$.

8. Let $z_0 \neq 0$ in $\mathbb{C}$. Prove that the function defined by the expression $|z|$ is not differentiable at z_0.

9. Determine whether or not there is any point in the complex plane at which the function defined by the expression $|z|^2$ is differentiable. Also decide whether or not $|z|$ is differentiable at 0.

10. Find a complex function that is differentiable everywhere in $\mathbb{C}$ and whose real part at $z = x + iy$ is $x^3 - 3xy^2 + e^x \cos y$.

11. Show that the union of every two open subsets of $\mathbb{C}$ is open, and that the intersection of every two open sets is open. Also extend both of these results from the case of *two* open sets to the more general case of any *finite* number of open sets.

12. Show that the union of every two closed subsets of $\mathbb{C}$ is closed, and that the intersection of every two closed sets is closed. Also extend both of these results from the case of *two* closed sets to the more general case of any *finite* number of closed sets.

13. For any family $\{G_\alpha : \alpha \in A\}$ of open subsets of $\mathbb{C}$ where the 'labelling set' A is infinite, show that the union

$$G = \cup_{\alpha \in A} G_\alpha = \{z \in \mathbb{C} : z \in G_\alpha \text{ for at least one } \alpha \in A\}$$

is also open.

14. For any family $\{K_\alpha : \alpha \in A\}$ of closed subsets of $\mathbb{C}$ where the 'labelling set' A is infinite, show that the intersection

$$K = \cap_{\alpha \in A} K_\alpha = \{z \in \mathbb{C} : z \in K_\alpha \text{ for every } \alpha \in A\}$$

is also closed.

15. Find a (necessarily infinite) family of open subsets of $\mathbb{C}$ whose intersection is not open, and a (necessarily infinite) family of closed subsets of $\mathbb{C}$ whose union is not closed.

16. (a) (Assuming the result of Theorem 4.4.2) given a complex function f that is regular on an open set G, prove that the real and imaginary parts of f are harmonic on G in the sense of Exercise 5 in Chapter 3. *Suggestion:* apply the Cauchy–Riemann equations both to f and to its derivative f'.

   (b) Is it possible for a function f to be differentiable everywhere in an open disc and have $x^5 + ax^3y^2 + bxy^4$ as its real part at every point $z = x + iy$ in that disc?

17. To throw further light onto paragraph 4.2.19:

(a) Find a 'decreasing' sequence $K_1 \supseteq K_2 \supseteq K_3 \supseteq K_4 \supseteq K_5 \cdots$ of non-empty bounded *open* subsets of $\mathbb{C}$ such that there is no point in common to all of them—that is, such that their intersection is empty.

(b) Find a 'decreasing' sequence $K_1 \supseteq K_2 \supseteq K_3 \supseteq K_4 \supseteq K_5 \cdots$ of non-empty *unbounded* closed subsets of $\mathbb{C}$ such that there is no point in common to all of them—that is, such that their intersection is empty.

18. Consider the complex function defined as follows:

$$
f(z) = \begin{cases} \dfrac{z^5}{|z|^4} & \text{if } z \neq 0, \\[2em] 0 & \text{if } z = 0. \end{cases}
$$

Verify that the real and imaginary parts of f do satisfy the Cauchy–Riemann equations at 0, but that f is not differentiable at 0.

19. Use sequence-based arguments to evaluate

$$
\lim_{z \to 3} \frac{z^4 - 81}{z^5 - 243} \quad \text{and} \quad \lim_{z \to i} \frac{z^6 + 1}{z^4 - 1}.
$$

20. Use epsilon-delta arguments to verify that

$$
\lim_{z \to 2+3i} (z^2 + 4\bar{z}) = 3 \quad \text{and} \quad \lim_{z \to 1+i} \frac{\text{Re}(z)}{\text{Im}(z)} = 1.
$$

21. Verify that, as $|z| \to \infty$:

$$
\frac{(\text{Re}(z) + \text{Im}(z))^2}{|z|^3} \to 0 \quad \text{and} \quad \frac{3z^2}{z^2 + 5z + 29} \to 3.
$$

22. **Introduction to complex logarithms**

(a) For any complex $z \neq 0$ verify that the solutions (for w in $\mathbb{C}$) of the equation $e^w = z$ are precisely the numbers $\log|z| + i \arg z$ where the notation indicates *any* of the argument angles of z, and log denotes the familiar real (natural) logarithm. These solutions are called the *(complex) logarithms of z.*

(b) The set of complex numbers $\mathbb{C} \setminus (-\infty, 0]$, that is, the entire complex plane apart from the non-positive real numbers, is known as the *cut plane*. (The intuition is that we have taken an impossibly sharp pair of scissors and cut out of the complex plane the whole of its negative real axis including zero.) This is the set on which Arg z, the principal value of the argument, lies in the *open* interval $(-\pi, \pi)$. Show that Arg z is a *continuous* function of z on $\mathbb{C} \setminus (-\infty, 0]$.

(c) Now we define a function Log : $\mathbb{C} \setminus (-\infty, 0] \to \mathbb{C}$ by the formula

$$\mathrm{Log}(z) = \log |z| + i \, \mathrm{Arg}(z), \quad z \in \mathbb{C} \setminus (-\infty, 0].$$

Briefly, why is Log continuous on $\mathbb{C} \setminus (-\infty, 0]$?
Log(z) as here described is called the *principal value* of log z or, sometimes, the *principal logarithm* of z.

(d) Prove that Log(z) is differentiable on the cut plane, and that its derivative there is $\dfrac{1}{z}$.

5 Background part C
Real integration

5.1 Introduction

Integration is the opposite of differentiation. That is,

- *'Determine an indefinite integral of the function f(x) over the interval J'*
  means
  go find (by any means whatsoever) another function F(x) whose derivative F′(x) equals f(x) so long as[1] x ∈ J.
- *'Evaluate the definite integral from a to b of f(x) (with respect to x)'*
  means
  find the function F(x) just referred to, and calculate its change-in-value F(b)−F(a) between a and b.

To be scrupulously honest, that opening paragraph is a jaw-dropping oversimplification. *Fully* to define a definite integral *from first principles* takes about half a dozen pages of thoughtful typing, and to establish that its elementary properties work as expected requires around three times that much careful argument, and that is still only the ground level of integration theory! Yet the culmination of that initial struggle—the so-called *fundamental theorem of calculus*—gives us all the tools that we normally use for actually calculating definite integrals, and that is really all we need by way of real-integration background to the main purpose of the present text (namely, investigating integration that is supported by *complex* numbers). So, very much in line with our approach to set theory back in Section 1.2, this is a huge and sophisticated area of study but, for our current needs, a fairly simple-minded view of some of its more elementary aspects will be perfectly adequate.

5.1.1 Definition

1. An *indefinite integral* of a real function $f(x)$ over an interval J means any function $F(x)$ whose derivative $F′(x)$ coincides with $f(x)$ on J. The symbol

---

[1] In point of fact we do not need $F′(x) = f(x)$ *at any endpoints* of J, so long as F is continuous at such endpoints.

Integration with Complex Numbers. McCluskey and McMaster, Oxford University Press.
© Brian McMaster and Aisling McCluskey (2022). DOI: 10.1093/oso/9780192846075.003.0005

$$\int f(x)\, dx$$

denotes (any or all of) these functions.

2. The *definite integral* of $f(x)$ from a to b (where $a < b$ in $\mathbb{R}$), usually denoted by

$$\int_a^b f(x)\, dx,$$

is the difference $F(b) - F(a)$ between the values at a and at b of the function F referred to in (1) above. This is often written as

$$[F(x)]_{x=a}^{x=b} \quad \text{or as} \quad [F(x)]_a^b.$$

This form of definition leaves unaddressed a number of questions such as

- How and when do we know that there is any function whose derivative is f?
- If there should happen to be more than one such function, how do we know that the definition of $\int_a^b f(x)\, dx$ gives the same answer irrespective of *which* of these functions we opt to work with?

Since this chapter seeks only to revise relevant background, we shall be content to state that

- If f is continuous on $[a, b]$ then there always does exist a function F such that $F' = f$ on that interval.
- There is, however, no guarantee that such a function can be explicitly described by a 'formula' in the normal sense of that word.
- The definition of the definite integral is, in all cases, independent of which indefinite integral we choose.
- At least in the case where $f(x)$ is positive on the interval $[a, b]$, the definite integral $\int_a^b f(x)\, dx$ has a useful geometrical interpretation: it is the area of the region lying between the graph of f, the horizontal axis, the vertical line $x = a$ and the vertical line $x = b$.

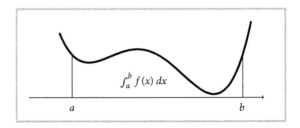

Table 5.1. Standard indefinite integrals

$f(x)$	$\int f(x)\,dx$	Condition?
x^a	$\dfrac{1}{a+1}x^{a+1}$	a any real constant except -1
$\cos x$	$\sin x$	
$\sin x$	$-\cos x$	
$\sec^2 x$	$\tan x$	x not of form $\frac{(2n+1)\pi}{2}, n \in \mathbb{Z}$
e^x	e^x	
$\dfrac{1}{x}$	$\log x$	$x > 0$
$\dfrac{1}{\sqrt{1-x^2}}$	$\arcsin x$	$-1 < x < 1$
$\dfrac{1}{1+x^2}$	$\arctan x$	

(Areas between graph and axis that lie *below* the axis need to be counted as negative if we want to extend this view to the case of non-positive functions.)

Our immediate objective is to remind the reader of some basic procedures for trying to find 'formulae' (but consider again the second bullet point above) for indefinite integrals of a variety of important functions.

5.2 Integration by inspection

Of course, the simplest way to integrate certain expressions (when it works) is merely to read Table 3.1 the other way around: please see Table 5.1, and note that there should be a '$+C$' in each entry of the second column because, whenever $F(x)$ is one function whose derivative is $f(x)$, then $F(x) + C$ (for any constant C) is another one.

Functions that closely resemble those in the first column of the table can also be integrated 'by inspection' provided that we take the trouble to differentiate an attempted answer to see if its derivative is as required, or is close enough to what is required that an improved attempt will get it exactly right. For example, if we are asked to integrate $\sec^2(\pi - 3x)$, row 4 of the table suggests an answer along the lines of 'the tangent of the same angle'. That is, $\tan(\pi - 3x)$ is a reasonable suggestion. If we carefully differentiate this (keeping the chain rule in mind), we find that its derivative is $-3\sec^2(\pi - 3x)$. So that's not quite right for our purposes, but it is only 'out by a numerical scale factor' since it is -3 times what we wanted. Therefore $-\frac{1}{3}$ times the earlier suggestion ought to be exactly right and, indeed, when we differentiate $-\frac{1}{3}\tan(\pi - 3x)$, we get a derivative of $\sec^2(\pi - 3x)$ as desired. That is,

$$\int \sec^2(\pi - 3x)\,dx = -\frac{1}{3}\tan(\pi - 3x) + C.$$

(Some of our students call this technique *integration by guessing*, and they are not altogether wrong.)

For another *by inspection* example, let us seek an indefinite integral for $\dfrac{1}{\sqrt{16 - x^2}}$ (with respect to x). It resembles row 7 of Table 5.1, except that the '16' does not match the '1'. If, however, we write

$$\frac{1}{\sqrt{16 - x^2}} \text{ as } \frac{1}{\sqrt{16\left(1 - \frac{x^2}{16}\right)}} = \frac{1}{4\sqrt{1 - \left(\frac{x}{4}\right)^2}}$$

then (temporarily ignoring the '4' on the bottom line since that is 'only a numerical scaling factor') it now resembles row 7 of Table 5.1 closely enough that we can consider (or guess!) a provisional answer of $\arcsin\left(\frac{x}{4}\right)$. Now, carefully differentiating the latter expression, we find that we get exactly the formula in the previous display. That is,

$$\int \frac{1}{\sqrt{16 - x^2}}\,dx = \arcsin\left(\frac{x}{4}\right) + C.$$

5.3 Integration by parts

The basic formula here[2]

$$\int f(x)g'(x)\,dx = f(x)g(x) - \int f'(x)g(x)\,dx$$

is really just a restatement of the product rule for differentiating

$$(f(x)g(x))' = f(x)g'(x) + f'(x)g(x).$$

It can in principle be used on any integrand that we happen to recognise as a product of **one function** times the derivative of **an other**, but will only help if

---

[2] Many will have memorised this in the abbreviated notation $\int uv' = uv - \int u'v$.

it then becomes easier to deal with '**the other function** times the derivative of **the one**'.

As a simple example, consider how to integrate (with respect to x) xe^{3x}. Here, either the x or the e^{3x} could be thought of as the derivative component of $f(x)g'(x)$. However, x is the derivative of the *slightly more complicated expression* $\frac{1}{2}x^2$, whereas e^{3x} is the derivative of the *no more complicated expression* $\frac{1}{3}e^{3x}$. Since it is hardly ever our ambition to make an integration question even marginally *more* complicated, the better strategy appears to be that in which we interpret the given function as $x(\frac{1}{3}e^{3x})'$. So the question transforms itself into

$$x\left(\frac{1}{3}e^{3x}\right) - \int (1)\left(\frac{1}{3}e^{3x}\right) dx,$$

which now unscrambles easily as $x(\frac{1}{3}e^{3x}) - \frac{1}{9}e^{3x} + C = \left(\dfrac{3x-1}{9}\right)e^{3x} + C$.

It may be timely to recall two further details:

1. Sometimes it is necessary to invoke integration by parts two or more times in succession within an exercise—see 5.3.1.

2. There are a small handful of integration problems (especially the integration of $\log x$ and of inverse trigonometric functions) that can be simplified by the apparently pointless action of multiplying the integrand by 1, but then treating that '1' as the derivative of x—see 5.3.2.

5.3.1 Example   Integrate $x^2 \sin x$ (with respect to x).

Solution   First, notice that

$$\int x \cos x \, dx = \int x(\sin x)' \, dx = x \sin x - \int (1)(\sin x) \, dx = x \sin x + \cos x.$$

Then, using this preliminary result,

$$\int x^2 \sin x \, dx = \int x^2(-\cos x)' \, dx = -x^2 \cos x - \int 2x(-\cos x) \, dx$$

$$= -x^2 \cos x + 2\int x \cos x \, dx = -x^2 \cos x + 2(x \sin x + \cos x)$$

$$= (2 - x^2)\cos x + 2x \sin x + C.$$

5.3.2 Example   Integrate $\log x$ (with respect to x).

Solution

$$\int \log x \, dx = \int \log x(1) \, dx = \int \log x(x)' \, dx = x \log x - \int x \left(\frac{1}{x}\right) \, dx$$

$$= x \log x - \int 1 \, dx = x \log x - x + C.$$

5.4 Integration by substitution, or change of variable

The basic formula for integration involving a substitution

$$\int f(x) \, dx = \int f(x(u)) \frac{dx}{du} \, du$$

is really just another way to express the chain rule, but in the context of integrating rather than differentiating. As in the case of the chain rule itself, it is commonly abbreviated into a form

$$\int y \, dx = \int y \frac{dx}{du} \, du$$

that makes it appear to be mere common sense, as if the 'number' du cancelled the bottom line of the 'fraction' $\frac{dx}{du}$ leaving the 'number' dx; but of course du and dx are not numbers (but, rather, indicators of which is the control variable) and $\frac{dx}{du}$ is not a fraction (but a derivative, defined as a limit). Nevertheless, the illusion of cancelling makes it easy to remember what this formula is saying.

5.4.1 Notes

1. Both chain rule and integration by substitution tell us that it is legitimate to change our minds as to which variable is in control, so long as we multiply by the appropriate 'conversion factor'. Unlike the chain rule, though, where it is usually fairly obvious in any given example *what* new variable to work with, integration by substitution sometimes offers us no explicit indication about which new variable would make the problem simpler.

2. By way of example, consider this integration challenge:

$$\int x^3 e^{x^4} \, dx.$$

It takes a certain amount of insight or luck or (more reliably) experience to see that setting $u = x^4$ will introduce a new variable that is effective in reducing this problem to a near-triviality. For then $\dfrac{du}{dx} = 4x^3$, that is, $x^3 = \dfrac{1}{4}\dfrac{du}{dx}$. So the challenge reshapes itself into

$$\frac{1}{4}\int \frac{du}{dx}e^u\, dx$$

which the basic rule translates as $\frac{1}{4}\int e^u\, du$, instantly recognisable as $\frac{1}{4}e^u$ ($+C$, of course). Converting back into terms of the original variable, we get an answer of $\frac{1}{4}e^{x^4} + C$.

3. A second example will serve as a reminder that, when integrating trigonometric expressions, it may be very useful to keep in mind the common trig identities. The problem

$$\int \sin^5 x \cos^3 x\, dx,$$

unlike the first, does not immediately present us with an integrand in which one part is (nearly) the derivative of another. Yet since $\cos^2 x = 1 - \sin^2 x$ identically, we can recast the question as

$$\int \sin^5 x \cos^2 x \cos x\, dx = \int \sin^5 x(1 - \sin^2 x)\cos x\, dx = \int(\sin^5 x - \sin^7 x)\cos x\, dx$$

at which point the substitution $u = \sin x$ (and $\dfrac{du}{dx} = \cos x$) transforms it into something easy:

$$\int(u^5 - u^7)\frac{du}{dx}\, dx = \int(u^5 - u^7)\, du = \frac{1}{6}u^6 - \frac{1}{8}u^8 = \frac{4\sin^6 x - 3\sin^8 x}{24} + C.$$

4. Here is another illustration of substitution that will turn out to be useful in Chapter 10: determine

$$\int \frac{x^2}{1 + x^6}\, dx.$$

(Roughwork... x^2 is not, of course, the derivative of x^6. Yet it is (almost) the derivative of x^3, and x^6 is the square of x^3. This suggests that x^3 is the key to this question.)

Put $u = x^3$ and observe that $\dfrac{1}{3}\dfrac{du}{dx} = x^2$. Then

$$\int \frac{x^2}{1 + x^6}\, dx = \frac{1}{3}\int \frac{1}{1 + u^2}\frac{du}{dx}\, dx = \frac{1}{3}\int \frac{1}{1 + u^2}\, du = \frac{1}{3}\arctan u = \frac{1}{3}\arctan(x^3) + C.$$

5.5 A look at improper integrals

A (definite) integral is said to be *improper* if either its domain is unbounded, or its integrand is undefined at one or more points of the domain, or both. A range of relatively simple examples will illustrate what that means in practice, and also indicate how to seek an interpretation—and indeed a numerical value—for certain improper integrals. For the sake of completeness we should define what is meant by saying that a real function $f(x)$ converges to a limit *as x tends to infinity*, although it is really only a slight adjustment of the 'complex' case covered by Definition 4.2.7.

5.5.1 Definition   Suppose that f is a (real) function whose domain includes arbitrarily large positive numbers, and that L is a (real) number. Then we say that $f(x)$ *tends to L as x tends to infinity* and write

$$f(x) \to L \text{ as } x \to \infty \quad \text{or} \quad \lim_{x \to \infty} f(x) = L$$

if, for each $\varepsilon > 0$, we can find some (large) positive number K_ε such that

$$x > K_\varepsilon \Rightarrow |f(x) - L| < \varepsilon.$$

(Also, $f(x) \to L$ as $x \to -\infty$ means that $f(x) \to L$ as $-x \to \infty$ or, equivalently, that $f(-x) \to L$ as $x \to \infty$.)

5.5.2 Example

1. The integral

$$\int_1^\infty x^{-2} \, dx$$

is improper because its domain is unbounded above. To seek a meaning for it, we imagine doing that integration *not* from 1 to infinity *but instead* from 1 to some greater number K (which we imagine to be very large). That part of the exercise, at least, is routine: $-x^{-1}$ has derivative equal to the integrand, so

$$\int_1^K x^{-2} \, dx = \left[-x^{-1} \right]_1^K = -\frac{1}{K} - \left(-\frac{1}{1} \right) = 1 - \frac{1}{K}.$$

Now when K is very big, that answer will be extremely close to 1: more precisely, the answer we obtained tends to a limit of 1 as K tends to infinity. We can express our findings by saying that this improper integral is *convergent*, and that

$$\int_1^\infty x^{-2} \, dx = 1.$$

2. The integral

$$\int_e^\infty x^{-1}\, dx$$

is improper because its domain is unbounded above. We again imagine doing that integration *not* from e to infinity *but instead* from e to some greater (possibly very great) number K. Now $\log x$ has derivative equal to the integrand, so

$$\int_e^K x^{-1}\, dx = [\log x]_e^K = \log K - \log e = -1 + \log K.$$

In contrast to item (1) above, when K is very big, that answer will be very big also and, in fact, not bounded above: more precisely, the answer we obtained does not have a limit as K tends to infinity. This represents a failure to assign any (finite) real value to the integral under examination: $\int_e^\infty x^{-1}\, dx$ is a *divergent* improper integral.

3. The integral

$$\int_{-\infty}^{-\frac{1}{\pi}} x^{-2} \sin(x^{-1})\, dx$$

is improper because its domain is unbounded below. This time we imagine integrating *not* from $-\infty$ to $-\frac{1}{\pi}$ *but instead* from some more negative (numerically very large) number $-K$ to $-\frac{1}{\pi}$. Now, following up from a change of variable (triggered by the fact that x^{-2} is almost exactly the derivative of x^{-1}) we find that $\cos(x^{-1})$ has derivative equal to the integrand, so

$$\int_{-K}^{-\frac{1}{\pi}} x^{-2} \sin(x^{-1})\, dx = \left[\cos(x^{-1})\right]_{-K}^{-\frac{1}{\pi}} = \cos(-\pi) - \cos\left(-\frac{1}{K}\right).$$

Since K (and $-K$) are huge, $-\frac{1}{K}$ will be very close to zero and (using continuity of cosine) its cosine will be virtually $\cos 0 = 1$: more precisely, our answer converges to $-1 - 1 = -2$ as $-K$ tends to minus infinity. We conclude that the improper integral is convergent, and that its numerical value is -2.

4. The integral

$$\int_0^1 \frac{1}{\sqrt{x}}\, dx$$

is improper because its integrand is undefined at 0, the leftmost point of the domain of integration. Here we consider instead the integral from h to 1, where

h is a (very small) positive number. Now the modified 'non-improper' integral is quite routine to evaluate:

$$\int_h^1 \frac{1}{\sqrt{x}}\, dx = \int_h^1 x^{-\frac{1}{2}}\, dx = \left[2x^{\frac{1}{2}}\right]_h^1 = 2\sqrt{1} - 2\sqrt{h}.$$

As h gets closer and closer to 0, so does $\sqrt{h}$; more precisely, as h tends to 0 while remaining positive, $2\sqrt{1} - 2\sqrt{h} \to 2$. We find that the integral is convergent and that

$$\int_0^1 \frac{1}{\sqrt{x}}\, dx = 2.$$

5. The integral

$$\int_{-1}^0 \frac{1}{x^2}\, dx$$

is improper because its integrand is undefined at 0, the rightmost point of the domain of integration. We look instead at the integral from -1 to h, where h is a (very small) negative number. The modified 'non-improper' integral is

$$\int_{-1}^h \frac{1}{x^2}\, dx = \left[-\frac{1}{x}\right]_{-1}^h = -\frac{1}{h} - \left(-\frac{1}{-1}\right) = -1 - \frac{1}{h}.$$

As h gets closer and closer to 0, this quantity becomes unboundedly large. (We could say that it diverges to infinity, but the main point is that it does not possess a limit.) The conclusion is that this improper integral is divergent.

We can now identify formally the pattern that is visible in these examples.

5.5.3 Definition   A definite integral is said to be *improper* if its domain (that is, its domain of integration) is unbounded or its integrand is undefined at a point of the domain. Of the four basic types:

1. An integral of the form $\int_a^\infty f(x)\, dx$ is called *convergent* if the limit

$$\lim_{K\to\infty} \int_a^K f(x)\, dx$$

exists, and *divergent* if it does not exist.

2. An integral of the form $\int_{-\infty}^b f(x)\, dx$ is called *convergent* if the limit

$$\lim_{K\to-\infty} \int_K^b f(x)\, dx \quad \text{or} \quad \lim_{K\to\infty} \int_{-K}^b f(x)\, dx$$

exists, and *divergent* if it does not exist.

3. An integral of the form $\int_a^b f(x)\,dx$ that is improper *because f(a) is undefined* is called *convergent* if the limit

$$\lim_{h>0,\ h\to 0} \int_{a+h}^{b} f(x)\,dx$$

exists, and *divergent* if it does not exist.

4. An integral of the form $\int_a^b f(x)\,dx$ that is improper *because f(b) is undefined* is called *convergent* if the limit

$$\lim_{h>0,\ h\to 0} \int_{a}^{b-h} f(x)\,dx$$

exists, and *divergent* if it does not exist.

An integral that is improper for more complicated reasons than these should be split up (if possible) into a finite number of improper integrals of types 1, 2, 3 and/or 4, and is then said to be *convergent* if all of these are convergent, and *divergent* if not all of them are convergent.

5.5.4 **Note**   The last sentence of that lengthy definition requires more explanation.

- How do we decide on the convergence or otherwise of an integral of the form $\int_{-\infty}^{\infty} f(x)\,dx$? We split it into $\int_{-\infty}^{0} f(x)\,dx$ and $\int_{0}^{\infty} f(x)\,dx$. If both of those converge, then the original integral is termed convergent (and is given a numerical value that is the sum of their two numerical values). Any other number than 0 could equally well be chosen as the 'splitting point', and this will not affect the outcome so long as there is no *other* reason why the integral was improper.

- How do we deal with an integral $\int_a^b f(x)\,dx$ where $f(c)$ fails to be defined at one *interior* point c of the domain, that is, such that $a < c < b$? We divide the problem into $\int_a^c f(x)\,dx$ and $\int_c^b f(x)\,dx$, which are improper of types 4 and 3, respectively. If both of those are convergent, then so is the original integral.

- How are we meant to seek an interpretation of $\int_{-\infty}^{\infty} \dfrac{\sin x}{x}\,dx$? The domain is unbounded in both directions, and there is also a point 0 at which the integrand is undefined. We can split the question as follows:

$$\int_{-\infty}^{-\pi} \frac{\sin x}{x}\,dx + \int_{-\pi}^{0} \frac{\sin x}{x}\,dx + \int_{0}^{\pi} \frac{\sin x}{x}\,dx + \int_{\pi}^{\infty} \frac{\sin x}{x}\,dx.$$

If all four of these 'basic' improper integrals converge, then so does the original. (There is nothing special about the numbers $-\pi$ and π in that decomposition, for any negative and any positive would have done just as well.)

There are various convergence tests for improper integrals. In general, they look very like analogous convergence tests for series, and they can be proved by much the same arguments. The following two will be particularly useful for us:

5.5.5 Lemma: 'absolute convergence implies convergence'   If the improper integral

$$\int_a^\infty |f(x)|\, dx$$

is convergent, then so is the improper integral

$$\int_a^\infty f(x)\, dx$$

(and similar results hold for the other basic types of improper integral).

5.5.6 Lemma: a 'direct comparison test'   If $0 \le f(x) \le g(x)$ for every x in $[a, \infty)$, and the improper integral

$$\int_a^\infty g(x)\, dx$$

is convergent, then so is the improper integral

$$\int_a^\infty f(x)\, dx$$

(and similar results hold for the other basic types of improper integral).

5.5.7 Example   Decide whether or not the improper integral

$$\int_1^\infty \frac{\sin x + e^{-x^2}}{x^2}\, dx$$

converges.

Solution   For all values of x in $[1, \infty)$ we have $|\sin x + e^{-x^2}| \le 1 + 1$ so, putting $f(x)$ = the given integrand and $g(x) = \dfrac{2}{x^2}$, we have $0 \le |f(x)| \le g(x)$ for all relevant x.

By a very slight modification of 5.5.2 (1) we see that $\int_1^\infty g(x)\, dx$ is convergent. Therefore so is $\int_1^\infty |f(x)|\, dx$ by 5.5.6.
Therefore so also is $\int_1^\infty f(x)\, dx$ by 5.5.5.

5.6 Cauchy principal values–a (slightly) more advanced topic

Some improper integrals, despite being divergent and therefore not possessing a numerical value in the sense that we discussed in Section 5.5, can have a numerical value *associated* with them if we use the symmetry present in their description as a means of reinterpreting them. This strikes most learners as a strange thing to attempt, so it may again be helpful to begin with a handful of examples rather than launching directly into the theory.

5.6.1 Example   Let us consider the two improper integrals

$$\int_{-\infty}^{\infty} \frac{2x}{1 + x^2}\, dx \text{ and}$$

$$\int_{-\infty}^{\infty} \frac{6x}{(1 + x^2)(4 + x^2)}\, dx.$$

Using the change of variable $u = 1 + x^2$, it is fairly easy to see that an *indefinite* integral for the first one is $\log(1 + x^2)$ (if in doubt, just use the chain rule to differentiate $\log(1 + x^2)$ and see that you get the first integrand). The second looks a bit harder, until you notice that the *partial fractions* trick splits the integrand into

$$\frac{2x}{1 + x^2} - \frac{2x}{4 + x^2}$$

for which an indefinite integral is $\log(1 + x^2) - \log(4 + x^2)$. Thanks to one of the standard rules of logarithms, this expression tidies into

$$\log\left(\frac{1 + x^2}{4 + x^2}\right).$$

Now we shall try three different approaches, *only one of which is correct according to Section 5.5*, in an attempt to evaluate the two definite integrals.

1. **A graphical view**
   Here are sketch graphs of the two integrands:

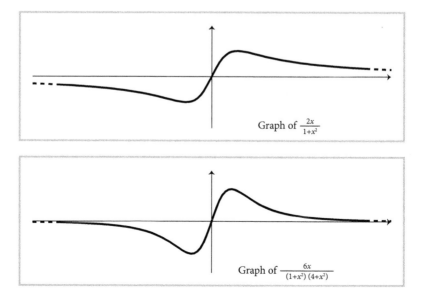

Graph of $\frac{2x}{1+x^2}$

Graph of $\frac{6x}{(1+x^2)(4+x^2)}$

It is visually obvious that each graph is symmetrical in the sense that, in each case, the region to the right of the origin between graph and horizontal axis exactly matches the region to the left of the origin between graph and horizontal axis—indeed, this is algebraically obvious also since both integrands are odd functions (that is, $f(-x) = -f(x)$ for all relevant x and each function f).

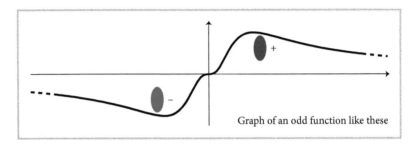

Graph of an odd function like these

More emphatically, every piece of area (under graph and above axis) that contributes positively towards the integral is precisely matched by another piece (between graph and axis but lying below the axis) that contributes negatively to it, and therefore cancels its contribution exactly. Therefore, in both cases, the graphical evidence suggests that the integral (thought of as area between graph and axis but counting areas below the axis as negative) 'ought to be zero'.

2. **A symmetrical limiting approach**
   It is easy to evaluate, for each (large) positive number K, the relevant integrals from $-K$ to K:

$$\int_{-K}^{K} \frac{2x}{1+x^2}\, dx = \left[\log(1+x^2)\right]_{-K}^{K} = \log(1+K^2) - \log(1+(-K)^2) = 0 \ \text{ and}$$

$$\int_{-K}^{K} \frac{6x}{(1+x^2)(4+x^2)}\, dx = \left[\log\left(\frac{1+x^2}{4+x^2}\right)\right]_{-K}^{K}$$

$$= \log\left(\frac{1+K^2}{4+K^2}\right) - \log\left(\frac{1+(-K)^2}{4+(-K)^2}\right) = 0.$$

Letting K tend to infinity in these, it is tempting (*even though it is incorrect!*) to conclude that

$$\int_{-\infty}^{\infty} \frac{2x}{1+x^2}\, dx = 0 \ \text{ and } \ \int_{-\infty}^{\infty} \frac{6x}{(1+x^2)(4+x^2)}\, dx = 0.$$

3. **Using Definition 5.5.3 (and Note 5.5.4)**
   Since the formal approach of 5.5.3 insists that we 'separate' the reasons for improperness in such integrations, the second integral should be divided up into

$$\int_{-\infty}^{0} \frac{6x}{(1+x^2)(4+x^2)} \, dx \; + \; \int_{0}^{\infty} \frac{6x}{(1+x^2)(4+x^2)} \, dx$$

each of which we can evaluate just as before:

$$\lim_{K \to \infty} \left[\log \left(\frac{1+x^2}{4+x^2} \right) \right]_{-K}^{0} \; + \; \lim_{K \to \infty} \left[\log \left(\frac{1+x^2}{4+x^2} \right) \right]_{0}^{K}$$

$$= \lim_{K \to \infty} \left(\log \frac{1}{4} - \log \left(\frac{1+(-K)^2}{4+(-K)^2} \right) \right) + \lim_{K \to \infty} \left(\log \left(\frac{1+K^2}{4+K^2} \right) - \log \frac{1}{4} \right)$$

noting that, under the strictures of 5.5.3, we must evaluate each limit separate-ly! Fortunately, as $K \to \infty$, $\dfrac{1+K^2}{4+K^2} = \dfrac{1/K^2+1}{4/K^2+1} \to \dfrac{1}{1} = 1$ so, since log is a continuous function, each of the two limits actually does exist, and

$$\lim_{K \to \infty} \left(\log \frac{1}{4} - \log \left(\frac{1+(-K)^2}{4+(-K)^2} \right) \right) = \log \frac{1}{4} - \log(1) = \log \frac{1}{4} \quad \text{and}$$

$$\lim_{K \to \infty} \left(\log \left(\frac{1+K^2}{4+K^2} \right) - \log \frac{1}{4} \right) = \log(1) - \log \frac{1}{4} = -\log \frac{1}{4}.$$

We conclude (and probably to no one's surprise) that $\displaystyle\int_{-\infty}^{\infty} \frac{6x}{(1+x^2)(4+x^2)} \, dx$ converges to $\log(1/4) - \log(1/4) = 0$.

However, when we use the same style of attack on the other integration question, something very different happens. We are again mandated by 5.5.3 to divide the integral into

$$\int_{-\infty}^{0} \frac{2x}{1+x^2} \, dx \; + \; \int_{0}^{\infty} \frac{2x}{1+x^2} \, dx$$

and work with the two limiting problems separately. This immediately runs into serious difficulty, because

$$\lim_{K \to \infty} \int_{-K}^{0} \frac{2x}{1+x^2} \, dx = \lim_{K \to \infty} \left[\log(1+x^2) \right]_{-K}^{0}$$

$$= \lim_{K \to \infty} (\log 1 - \log(1+K^2)) = - \lim_{K \to \infty} \log(1+K^2)$$

does not exist (and, likewise, the integral from 0 to K does not have a limit as $K \to \infty$). This integral then, despite the more-or-less plausible arguments put forward for its convergence (to zero) in the first and second approaches above, is in reality a *divergent* improper integral.

For those readers who have time and curiosity enough, here in outline is a second pair of examples, this time focusing on integrals that are improper because the integrand fails to be defined at an interior point of the domain of integration. We shall again seek insight into the effect of approaching the point of difficulty *symmetrically*.

5.6.2 **Example**   We consider the improper integrals

$$\int_{-1}^{2} \frac{2}{5}x^{-\frac{3}{5}}\, dx \quad \text{and} \quad \int_{-1}^{2} \frac{2}{5}x^{-\frac{7}{5}}\, dx,$$

whose integrands fail definition at $x = 0$. Clearly, their indefinite integrals (so long as $x \neq 0$) are $x^{\frac{2}{5}}$ and $-x^{-\frac{2}{5}}$.

1. In both cases, sketch graphs of the integrands look approximately like this:

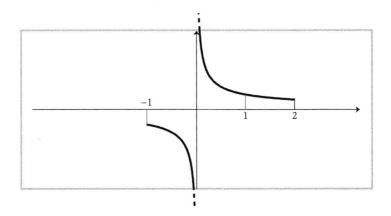

   and they *suggest* that the area between graph and axis that lies below the axis precisely matches (and therefore cancels?) part of the area that lies above the axis, leaving us with only the integral from 1 to 2. That evaluates to $2^{\frac{2}{5}} - 1$ in the first case and $1 - 2^{-\frac{2}{5}}$ in the second.

2. The way to approach this kind of singularity symmetrically is to think of a small positive number h, evaluate and add the integrals from -1 to $-h$ and from h to 2, and then seek a limit of that answer as $h \to 0$. This works quite routinely on both of the current examples:

$$\int_{-1}^{-h} \frac{2}{5}x^{-\frac{3}{5}}\, dx + \int_{h}^{2} \frac{2}{5}x^{-\frac{3}{5}}\, dx = [x^{\frac{2}{5}}]_{-1}^{-h} + [x^{\frac{2}{5}}]_{h}^{2}$$

$$= h^{\frac{2}{5}} - 1 + 2^{\frac{2}{5}} - h^{\frac{2}{5}} \to 2^{\frac{2}{5}} - 1 \text{ as } h \to 0, \quad \text{and}$$

$$\int_{-1}^{-h} \frac{2}{5}x^{-\frac{7}{5}}\, dx + \int_{h}^{2} \frac{2}{5}x^{-\frac{7}{5}}\, dx = [-x^{-\frac{2}{5}}]_{-1}^{-h} + [-x^{-\frac{2}{5}}]_{h}^{2}$$

$$= -h^{-\frac{2}{5}} + 1 - 2^{-\frac{2}{5}} + h^{-\frac{2}{5}} \to 1 - 2^{-\frac{2}{5}} \text{ as } h \to 0.$$

3. To explore properly the convergence of these integrals using 5.5.3, we need to investigate separately the convergence (or otherwise) of each integral from -1 to 0 and from 0 to 2. In the case of the function $f(x) = \frac{2}{5}x^{-\frac{3}{5}}$, and picking up the story from our discussion in (2) above, that entails looking for limits (as $h \to 0$) of $h^{\frac{2}{5}} - 1$ and of $2^{\frac{2}{5}} - h^{\frac{2}{5}}$ before putting them together. This works smooth-ly because these two limits both exist: they evaluate as -1 and $2^{\frac{2}{5}}$. Therefore $\int_{-1}^{2} f(x)\, dx$ exists (converges) and equals $2^{\frac{2}{5}} - 1$ (in agreement with both (1)'s and (2)'s attempts at its evaluation).

   On the other hand, in the case of $g(x) = \frac{2}{5}x^{-\frac{7}{5}}$, the attempts to evaluate the limits (as $h \to 0$) of $-h^{-\frac{2}{5}} + 1$ and of $-2^{-\frac{2}{5}} + h^{-\frac{2}{5}}$ both fail—neither limit exists, that is, neither the integral from -1 to 0 nor the integral from 0 to 2 of the function g converges. Thus $\int_{-1}^{2} g(x)\, dx$ is a divergent improper integral, *and the arguments put forward in (1) and (2) above for its evaluation were both incorrect.*

   Although what we have called the 'symmetrical approach' to exploring improp-er integrals of these kinds has been shown above to be potentially misleading (or you might even opt for the less equivocal term *wrong!*) it is not without significance within and beyond our present ambitions. The name generally given to a symmet-rically obtained limit in such circumstances is *Cauchy principal value*, sometimes abbreviated to CPV.

5.6.3 Definition

- If the limit

$$\lim_{K \to \infty} \int_{-K}^{K} f(x)\, dx$$

  exists, then it is called the *Cauchy principal value* of the improper integral $\int_{-\infty}^{\infty} f(x)\, dx$, sometimes denoted by CPV $\int_{-\infty}^{\infty} f(x)\, dx$.

- Suppose that the integral $\int_{a}^{b} f(x)\, dx$ is improper because the integrand f is unde-fined at a single point c such that $a < c < b$. If the limit

$$\lim_{h > 0, \, h \to 0} \left(\int_{a}^{c-h} f(x)\, dx + \int_{c+h}^{b} f(x)\, dx \right)$$

  exists, then it is called the *Cauchy principal value* of the improper integral $\int_{a}^{b} f(x)\, dx$, sometimes denoted by CPV $\int_{a}^{b} f(x)\, dx$.

Since this material is likely to be unfamiliar to many readers, we should devote a little time to exploring a further couple of examples.

5.6.4 **Example**   For the improper integrals

$$\int_{-\infty}^{\infty} \frac{x^3}{81 + x^4}\, dx \quad \text{and} \quad \int_{-\infty}^{\infty} \frac{x}{81 + x^4}\, dx,$$

determine the Cauchy principal values. Also decide whether each integral is actually convergent.

Solution   The derivative of $\log(81+x^4)$ is $\dfrac{4x^3}{81 + x^4}$, and the derivative of $\arctan(x^2/9)$ is $\dfrac{18x}{81 + x^4}$. Therefore, for each $K > 0$:

$$\int_{-K}^{K} \frac{x^3}{81 + x^4}\, dx = \left[\frac{1}{4}\log(81 + x^4)\right]_{-K}^{K} = 0, \text{ and}$$

$$\int_{-K}^{K} \frac{x}{81 + x^4}\, dx = \left[\frac{1}{18}\arctan(x^2/9)\right]_{-K}^{K} = 0.$$

Taking limits as $K \to \infty$, we get

$$\text{CPV} \int_{-\infty}^{\infty} \frac{x^3}{81 + x^4}\, dx = 0 \quad \text{and} \quad \text{CPV} \int_{-\infty}^{\infty} \frac{x}{81 + x^4}\, dx = 0.$$

Yet also

$$\int_{0}^{K} \frac{x^3}{81 + x^4}\, dx = \left[\frac{1}{4}\log(81 + x^4)\right]_{0}^{K} = \frac{1}{4}\log(81 + K^4) - \log 3$$

which is unbounded as $K \to \infty$, and similar remarks are valid for the integral from $-K$ to 0. So all of the integrals

$$\int_{0}^{\infty} \frac{x^3}{81 + x^4}\, dx \qquad \int_{-\infty}^{0} \frac{x^3}{81 + x^4}\, dx \qquad \int_{-\infty}^{\infty} \frac{x^3}{81 + x^4}\, dx$$

are actually divergent.

On the other hand,

$$\int_0^K \frac{x}{81 + x^4} \, dx = \frac{1}{18} \arctan(K^2/9) - \frac{1}{18} \arctan 0$$

which converges to $\dfrac{\pi}{36}$ as $K \to \infty$, and a similar analysis shows that the integral from $-K$ to 0 converges to $-\dfrac{\pi}{36}$. So the integrals

$$\int_0^\infty \frac{x}{81 + x^4} \, dx \qquad \int_{-\infty}^0 \frac{x}{81 + x^4} \, dx \qquad \int_{-\infty}^\infty \frac{x}{81 + x^4} \, dx$$

all converge, and their (convergent) values are $\dfrac{\pi}{36}, -\dfrac{\pi}{36}$ and 0.

5.6.5 Example Determine the Cauchy principal values of the improper integrals

$$\int_{-8}^{27} \frac{1}{\sqrt[3]{x}} \, dx \quad \text{and} \quad \int_{-2}^3 \frac{1}{x^3} \, dx.$$

Also decide whether each integral is convergent or divergent.

Solution The derivative of $x^{2/3}$ is $(2/3)(1/\sqrt[3]{x})$, and the derivative of x^{-2} is $-2x^{-3}$. Therefore, for each small positive number h,

$$\int_{-8}^{-h} \frac{1}{\sqrt[3]{x}} \, dx + \int_h^{27} \frac{1}{\sqrt[3]{x}} \, dx = \left[\frac{3}{2} x^{\frac{2}{3}}\right]_{-8}^{-h} + \left[\frac{3}{2} x^{\frac{2}{3}}\right]_h^{27} = \frac{15}{2}, \text{ and}$$

$$\int_{-2}^{-h} \frac{1}{x^3} \, dx + \int_h^3 \frac{1}{x^3} \, dx = \left[-\frac{1}{2}\frac{1}{x^2}\right]_{-2}^{-h} + \left[-\frac{1}{2}\frac{1}{x^2}\right]_h^3 = \frac{5}{72}.$$

Taking limits as $h \to 0$, we get

$$CPV \int_{-8}^{27} \frac{1}{\sqrt[3]{x}} \, dx = \frac{15}{2} \quad \text{and} \quad CPV \int_{-2}^3 \frac{1}{x^3} \, dx = \frac{5}{72}.$$

Yet also

$$\int_h^3 \frac{1}{x^3} \, dx = \left[-\frac{1}{2}\frac{1}{x^2}\right]_h^3 = -\frac{1}{18} + \frac{1}{2h^2}$$

which is unbounded as $h \to 0$, and similar remarks are valid for the integral from -2 to $-h$. So all of the integrals

$$\int_{-2}^0 \frac{1}{x^3} \, dx \qquad \int_0^3 \frac{1}{x^3} \, dx \qquad \int_{-2}^3 \frac{1}{x^3} \, dx$$

are actually divergent.

On the other hand,

$$\int_{h}^{27} \frac{1}{\sqrt[3]{x}} \, dx = \left[\frac{3}{2} x^{\frac{2}{3}} \right]_{h}^{27} = \frac{27}{2} - \frac{3}{2} h^{\frac{2}{3}}$$

which converges to $\dfrac{27}{2}$ as $h \to 0$, and a similar analysis shows that the integral from -8 to $-h$ converges to -6. So the integrals

$$\int_{-8}^{0} \frac{1}{\sqrt[3]{x}} \, dx \qquad \int_{0}^{27} \frac{1}{\sqrt[3]{x}} \, dx \qquad \int_{-8}^{27} \frac{1}{\sqrt[3]{x}} \, dx$$

all converge, and their (convergent) values are -6, $\dfrac{27}{2}$ and $\dfrac{15}{2}$.

5.6.6 **Proposition**   If an improper integral of the form

$$\int_{-\infty}^{\infty} f(x) \, dx$$

is convergent, then its (convergent) numerical value coincides with its Cauchy principal value. (In less formal terms, if the integral actually converges, then the CPV shortcut 'gets it right'.)

Proof   Suppose that the integral converges. Then (by definition) as $K \to \infty$:

$$\int_{0}^{K} f(x) \, dx \to \int_{0}^{\infty} f(x) \, dx, \qquad \int_{-K}^{0} f(x) \, dx \to \int_{-\infty}^{0} f(x) \, dx, \quad \text{and}$$

$$\int_{-\infty}^{\infty} f(x) \, dx = \int_{-\infty}^{0} f(x) \, dx + \int_{0}^{\infty} f(x) \, dx.$$

To seek the CPV, we take a large positive number K, and look at

$$\int_{-K}^{K} f(x) \, dx = \int_{-K}^{0} f(x) \, dx + \int_{0}^{K} f(x) \, dx$$

which tends to

$$\int_{-\infty}^{0} f(x) \, dx + \int_{0}^{\infty} f(x) \, dx = \int_{-\infty}^{\infty} f(x) \, dx$$

as required.   ■

Our reasons for including a section on Cauchy principal values of improper integrals will not fully come to light until Chapter 10. They are probably far from obvious to most readers at this stage, so we ought to outline them before concluding the current chapter.

- One of the central messages this text wants to communicate is that many apparently difficult real integral problems can be solved, and solved relatively easily and quickly, by converting them into complex integral problems for which powerful and elegant methods are available.
- Many of these 'difficult real integral problems' are actually improper integrals.
- However, when we use complex integration techniques on an improper real integral, what they usually deliver is its Cauchy principal value.
- That's the end of the problem *provided we know that the improper integral is actually convergent*: because, as is easy to prove,[3] the Cauchy principal value of a *convergent* integral coincides with its (properly defined) convergent numerical value. (Remember also that there are tests such as 5.5.5 and 5.5.6 that can be employed to check on convergence of improper integrals.)
- Even when the original (real) improper integral is *divergent*, its Cauchy principal value is still a potentially useful finding. Applications of this, however, lie beyond the scope of the present text.
- When we determine the Cauchy principal value of a divergent improper integral, it is important to be aware that this is all we have done. For instance, if, in 5.6.1 (2), after confirming that $\int_{-K}^{K} \frac{2x}{1 + x^2} \, dx$ possessed a limit of 0 as $K \to \infty$, we had overstated our conclusion by claiming that $\int_{-\infty}^{\infty} \frac{2x}{1 + x^2} \, dx$ was convergent to 0, we would have been guilty of a mathematical heresy similar to claiming that 'infinity minus infinity equals zero'! An awareness of Cauchy principal value, and especially of its limitations, is therefore an important notion to take on board.

5.7 **Exercises**

1. Verify that the improper integral

$$\int_{-1}^{2} \frac{1}{|x|^t} \, dx$$

   is convergent if $0 < t < 1$ (and evaluate it), but divergent if $t \geq 1$.

2. Do the following integrals converge or diverge?

$$\int_{0}^{\infty} \frac{4(x^2 - 1)}{x^4 + x^2 + 1} \, dx, \quad \int_{-\infty}^{0} \frac{4(x^2 - 1)}{x^4 + x^2 + 1} \, dx, \quad \int_{-\infty}^{\infty} \frac{4(x^2 - 1)}{x^4 + x^2 + 1} \, dx.$$

   In each case of convergence, determine the integral's value.

---

[3] See 5.6.6, for example.

3. Show that
$$\int_{-\infty}^{\infty} \frac{x^2 - x + 2}{x^4 + 10x^2 + 9} \, dx$$
is convergent, and evaluate it.

4. (a) Show that
$$\int_{0}^{\infty} \frac{3(x-1)}{x^3 + 1} \, dx$$
is convergent, and evaluate it.

   (b) Show that
$$\int_{-\infty}^{\infty} \frac{3(x-1)}{x^3 + 1} \, dx$$
is divergent, and evaluate its CPV.

5. Show that each of the following is divergent, and determine its CPV:

   (a)
$$\int_{-\infty}^{\infty} \frac{4x^3}{x^4 + 1} \, dx,$$

   (b)
$$\int_{0}^{2} \frac{4x^3}{x^4 - 1} \, dx,$$

   (c)
$$\int_{-3}^{2} \frac{4x^3}{x^4 - 1} \, dx.$$

6. (a) If the real function $F : [a, \infty) \to \mathbb{R}$ is increasing, that is, if $x \le y \Rightarrow F(x) \le F(y)$, show that $F(x)$ converges to some limit as $x \to \infty$ if and only if F is bounded above on $[a, \infty)$.

   (b) Prove 5.5.6: that if $0 \le f(x) \le g(x)$ on $[a, \infty)$ and $\int_a^\infty g(x) \, dx$ is convergent, then also $\int_a^\infty f(x) \, dx$ is convergent.

7. Given that a is a real constant such that $0 < a < 1$, prove that the following improper integral is convergent:
$$\int_{0}^{\infty} \frac{x^{a-1}}{x + 1} \, dx.$$

8. If $f : [a, b) \cup (b, c] \to \mathbb{R}$ and $f(b)$ is not defined and both $\int_a^b f(x) \, dx$ and $\int_b^c f(x) \, dx$ are convergent, show that the Cauchy principal value of $\int_a^c f(x) \, dx$ agrees with the (convergent) value of that integral.

9. *We recommend not trying the following problem until you have some experience of working with Cauchy sequences and, in particular, with the fact that a sequence is Cauchy if and only if it is convergent.*

(a) Show that a real function $F : [a, \infty) \to \mathbb{R}$ possesses a limit as $x \to \infty$ if and only if

$$\forall \varepsilon > 0 \ \exists \, x_\varepsilon \in [a, \infty) \text{ such that for all } x, y \geq x_\varepsilon, \ |F(x) - F(y)| < \varepsilon.$$

(b) Prove that if

$$\int_a^\infty |f(x)| \, dx$$

is convergent, then so is

$$\int_a^\infty f(x) \, dx.$$

6 Paths in the complex plane

6.1 Introduction

When we seek to evaluate a *real* definite integral, say, the integral

$$\int_a^b f(t)\, dt$$

of a real function f from a to b, then[1] the behaviour of $f(t)$ when t is less than a or greater than b is completely irrelevant; all that matters is how $f(t)$ behaves while t lies within the interval $[a, b]$. Although the phrase *from a to b* gives an impression that the value of t changes or moves during the integration process, the question of *how it moves* has such an obvious answer (*it starts at a and gradually increases until it reaches b*) that it is seldom even asked.

Now suppose we try to formulate as similar as possible a question in *complex* integration, for instance, to integrate the complex function

$$f(z) = \frac{1}{z}$$

from $z_0 = 2i$ to $z_1 = 1 - \sqrt{3}i$. We immediately face the issue of *how z is meant to change from z_0 to z_1*, and it no longer has an obvious default solution. Is the intention to base our answer on the values of $f(z)$ while z is restricted to the straight line joining z_0 to z_1? Or perhaps to the angled line joining z_0 to 0 and from there to z_1? Or the right-angled line from z_0 to $1 + 2i$ and thence to z_1? Or perhaps along an arc of the circle centre 0 radius 2 that z_0 and z_1 both lie on? And over all this pondering hangs the broader question *does it actually matter how z moves, or is the answer independent of our choice of 'pathway'?*

---

[1] We shall assume here that $a < b$.

Integration with Complex Numbers. McCluskey and McMaster, Oxford University Press.
© Brian McMaster and Aisling McCluskey (2022). DOI: 10.1093/oso/9780192846075.003.0006

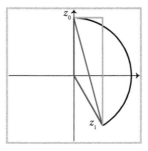

We can answer a small part of that set of questions immediately since $f(z) = \frac{1}{z}$ is undefined when $z = 0$: therefore, the second proposed pathway that was suggested above passes directly through a point 0 at which f fails to be defined. This choice of pathway would certainly be problematic for trying to evaluate the integral and, to at least this extent, we cannot expect the outcome to be independent of *how we imagine travelling from z_0 to z_1 as the integral develops.*

Yet there is a much greater dependence of the integral from z_0 to z_1 upon how we choose to travel from one point to the other, and it may be useful to admit that right away (at the risk of *spoilers*). As soon as we have properly defined the integral of a complex function along a pathway (which is actually quite easy once we get clear as to what all that ought to mean) then it will be simple to show that

- when we choose the pathway from z_0 to z_1 that follows the shorter arc of the circle centre 0 radius 2, then $\int_{z_0}^{z_1} \frac{1}{z}\, dz$ is approximately $-2.618i$,

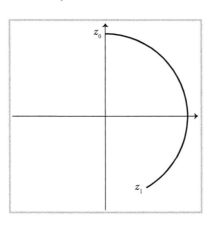

- but when instead we follow the longer arc, then $\int_{z_0}^{z_1} \frac{1}{z}\, dz$ is approximately $3.665i$.

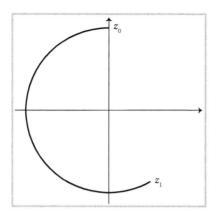

What this example (and many others like it) will show us is that the desire to describe 'the integral of a complex function from one location in the complex plane to another' actually misses the point, because the answer will depend on how we choose to travel. What we need to do instead is to define integrals *along a particular pathway*. Evidently, the first step in that process must be to make clear what the so-far ill-defined term 'pathway' ought to mean in order that the integrals then defined shall work properly.

6.2 Functions from $\mathbb{R}$ to $\mathbb{C}$

Our definition of pathway is going to be delivered by functions whose domain is a set of *real* numbers but whose output values are *complex* numbers, and we shall require these pathway functions to be at the very least continuous, and usually also differentiable, and to take part in integration processes. For this purpose we now need to extend the basic calculus notions that we have seen developed for real functions, and that we have already extended to complex functions, so as to be able to use them again for these hybrid, 'real-to-complex' functions. The good news is that this extension is entirely routine: the definitions and results (and even the proofs) are just what you would expect on the basis of earlier chapters.

6.2.1 Definition   Let A be a subset of $\mathbb{R}$, and $\gamma : A \to \mathbb{C}$ a complex-valued function on A. We say that γ is *continuous at a point p of A* if the following equivalent conditions are satisfied:

- For each sequence (t_n) in A such that $t_n \to p$ (as $n \to \infty$) we have $\gamma(t_n) \to \gamma(p)$.
- For each positive real number ε there is a positive real number δ_ε such that $|t - p| < \delta_\varepsilon$ and $t \in A$ together imply that $|\gamma(t) - \gamma(p)| < \varepsilon$.

In the case where γ is continuous at every point of its domain A, we call it a *continuous function*.

Any such function can, if desired, be separated out into its real and imaginary parts by writing $\gamma(t) = \text{Re}(\gamma(t)) + i\,\text{Im}(\gamma(t))$ for each $t \in A$, and such a separation neither creates nor destroys continuity:

6.2.2 Lemma   In 6.2.1's notation, $\gamma(t)$ is continuous if and only if both $\text{Re}(\gamma(t))$ and $\text{Im}(\gamma(t))$ are continuous.

6.2.3 Definition   Given $\gamma : A \to \mathbb{C}$ with $A \subseteq \mathbb{R}$, and an *interior point p* of A (that is, we can find some little open interval I centred on p so that $I \subseteq A$), we say that γ is *differentiable* at p if the limit

$$\lim_{h \to 0} \frac{\gamma(p + h) - \gamma(p)}{h} \quad \text{or, equivalently,} \quad \lim_{t \to p} \frac{\gamma(t) - \gamma(p)}{t - p}$$

exists, and this limit is then called the *derivative* of γ at p, usually written as $\gamma'(p)$.

6.2.4 Lemma   In 6.2.3's notation, γ is differentiable at p if and only if both $\text{Re}(\gamma(t))$ and $\text{Im}(\gamma(t))$ are differentiable at $t = p$. Furthermore, $\gamma'(p)$ is then $(\text{Re}(\gamma(t)))' + i(\text{Im}(\gamma(t)))'$ evaluated at $t = p$.

6.2.5 Note   In the case where A is an open interval in $\mathbb{R}$, a function $\gamma : A \to \mathbb{C}$ that is differentiable at every point of A is simply called *differentiable* or, if the emphasis seems necessary, *differentiable on A*.

In the case where A is a *closed* interval, say, $A = [a, b]$, we run into the usual problem about differentiating at endpoints, and we solve it as before by resorting to using one-sided limits at a and at b:

If $\gamma : [a, b] \to \mathbb{C}$ then

1. γ is said to be *differentiable at a* if

$$\lim_{h > 0, h \to 0} \frac{\gamma(a + h) - \gamma(a)}{h} \quad \text{or, equivalently,} \quad \lim_{t > a, t \to a} \frac{\gamma(t) - \gamma(a)}{t - a}$$

exists, and this limit is then usually still written as $\gamma'(a)$,

2. γ is said to be *differentiable at b* if

$$\lim_{h < 0, h \to 0} \frac{\gamma(b + h) - \gamma(b)}{h} \quad \text{or, equivalently,} \quad \lim_{t < b, t \to b} \frac{\gamma(t) - \gamma(b)}{t - b}$$

exists, and this limit is then usually still written as $\gamma'(b)$,

3. γ is said to be *differentiable on $[a, b]$* if it is differentiable on (a, b) in the usual sense, and also differentiable both at a and at b in the one-sided sense of the present note.

Defining the integral of a complex-valued function *of a real variable* presents no problem at all: one simply does the real and imaginary parts separately:

6.2.6 Definition   Let $\gamma : [a, b] \to \mathbb{C}$. We define

$$\int_a^b \gamma(t)\, dt = \int_a^b \mathrm{Re}(\gamma(t))\, dt + i \int_a^b \mathrm{Im}(\gamma(t))\, dt$$

provided that both of these (real) integrals exist. (They will if the integrands are continuous, of course.)

6.2.7 Exercise   Show how to evaluate $\int_0^\pi e^{it}(t + i \sin t)\, dt$.

Partial solution   The (complex-valued) function $\gamma(t) = e^{it}(t + i \sin t)$ expands out as

$$(\cos t + i \sin t)(t + i \sin t) = t \cos t - \sin^2 t + i(t \sin t + \sin t \cos t)$$

from which we can read off its real and imaginary parts. The problem has now become one of basic real calculus: to evaluate the two real integrals

$$a = \int_0^\pi (t \cos t - \sin^2 t)\, dt \ \text{ and } \ b = \int_0^\pi (t \sin t + \sin t \cos t)\, dt$$

after which the answer will be $a + ib$.

The basic results about integrals of complex-valued functions of a real variable that you would expect—for instance, that $\int_a^b (\gamma(t) + \beta(t))\, dt = \int_a^b \gamma(t)\, dt + \int_a^b \beta(t)\, dt$ and that $\int_a^b (K\gamma(t))\, dt = K \int_a^b \gamma(t)\, dt$ for any constant K—follow routinely from the definition. One that is less obvious, and which we shall need later on, is the following:

6.2.8 Theorem   If $\gamma : [a, b] \to \mathbb{C}$ is continuous, then

$$\left| \int_a^b \gamma(t)\, dt \right| \le \int_a^b |\gamma(t)|\, dt.$$

Proof   Express $\int_a^b \gamma(t)\, dt$ in modulus-argument form as $Re^{i\theta}$. Our task is to estimate R. Now

$$R = e^{-i\theta} \int_a^b \gamma(t)\, dt = \int_a^b e^{-i\theta} \gamma(t)\, dt.$$

Also R is a real number, so

$$R = \mathrm{Re}\left(\int_a^b e^{-i\theta} \gamma(t)\, dt \right)$$

$$= \int_a^b \mathrm{Re}\left(e^{-i\theta}\gamma(t)\right)\,dt \quad \text{(since that is how we integrate such functions)}$$

$$\leq \int_a^b \left|e^{-i\theta}\gamma(t)\right|\,dt \quad \text{(since real part cannot exceed modulus,}$$

and real integration respects inequalities)

$$= \int_a^b |\gamma(t)|\,dt \quad \text{as required.}$$

∎

6.3 Paths and contours

6.3.1 Definitions

1. A *path* in the complex plane $\mathbb{C}$ is[2] a continuous mapping

$$\gamma : [a, b] \to \mathbb{C}$$

   defined on a closed, bounded interval of the real line $\mathbb{R}$. The complex numbers $\gamma(a)$ and $\gamma(b)$ are its *first point* or *start*, and its *last point* or *end*. The path γ is often referred to as a *path from $\gamma(a)$ to $\gamma(b)$*.

2. The set of complex numbers $\{\gamma(t) : t \in [a, b]\}$ (which, for general mappings, would usually be called the *range* of γ) is known as the *track* of γ. If G is a subset of $\mathbb{C}$ that happens to contain the track of γ, then we sometimes refer to γ as a *path in G*.

3. We call the path γ a *closed path* if $\gamma(a) = \gamma(b)$ (that is, if its first and last points coincide).

4. We call the path γ *simple* if, for $t_1 \neq t_2$ in $[a, b]$, it never happens that $\gamma(t_1) = \gamma(t_2)$ except, possibly, when t_1 and t_2 are a and b (in other terminology, if the restriction of γ to $[a, b)$ is a one-to-one function).

6.3.2 Notes   (This paragraph is not, strictly speaking, mathematics, but it provides a way of thinking and talking about paths that many people find helpful and supportive of intuition.)

1. A path $\gamma : [a, b] \to \mathbb{C}$ can be thought of as the history of a moving particle, initially at the first point $\gamma(a)$, which travels continuously until eventually reaching the last point $\gamma(b)$. Its track is the set of all positions that it passes through. It is closed if it ultimately returns to its starting point. It is simple if the track never goes through the same point twice (except, maybe, that the start and end could

---

[2] There is a good deal of variation between different writers in the use of terms such as *path, line, curve, arc, contour etc.*

be the same point). In this mindset, $[a, b]$ is thought of as a *time interval*, and $\gamma(t)$ is the position that the particle is passing through at time t.

2. Two paths $\gamma : [a, b] \to \mathbb{C}$ and $\beta : [c, d] \to \mathbb{C}$ are regarded as equivalent if they have the same track, the same first point, the same last point, and they pass through the points on their track in the same order. In this connection, notice that

(a) A straight-line path from z_0 to z_1 and a straight-line path from z_1 to z_0 are not equivalent (in this sense) even though they have the same track: for the first point of the former is the last point of the latter and *vice versa*, and their points are certainly not passed through in the same order. The arrowheads in the diagrams are trying to make this distinction visible.

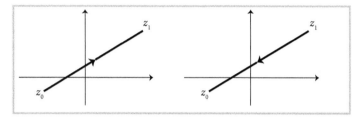

(b) A circular path from a particular point *anticlockwise* through a complete revolution and returning back to its starting point (we shall show a little later how properly to formulate paths like this as functions on a closed bounded real interval) is not equivalent to the same circle and same starting and ending point traced *clockwise* even though they have the same track, the same first point and the same last point: for they do not pass through points in the same order. The arrows in the diagrams are attempting to draw our attention to this.

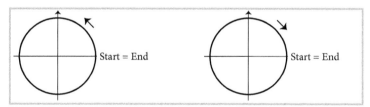

(c) Consider a first path formed by straight segments from $(0, 0)$ to $(1, 1)$, then from $(1, 1)$ to $(2, 1)$, then from $(2, 1)$ to $(3, 0)$.

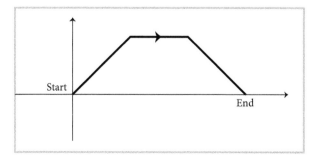

Consider also a second path formed by straight segments from $(0,0)$ to $(1,1)$, then from $(1,1)$ to $(2,1)$, then from $(2,1)$ to $(1,1)$, then from $(1,1)$ to $(2,1)$ and finally from $(2,1)$ to $(3,0)$.

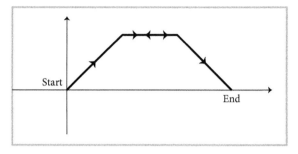

Once again, although possessing identical tracks, identical first points and identical last points, the paths are not equivalent. At this stage, the arrowheads are not entirely helpful in making that distinction clear, and we shall evidently need to pay increasing attention to the functions that define these paths rather than only to the sketches that try to render their behaviour visible.

(d) A circular path transcribed *three times* anticlockwise is not equivalent to either of those single-orbit circles mentioned earlier, and nor is a circular path transcribed (let us say) *five times* clockwise. In such cases, diagrammatic representations with the right number of arrowheads are of rather doubtful value!

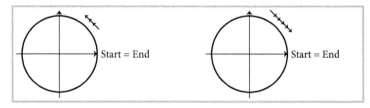

3. It is legitimate, but not obligatory, to assume that all paths have the domain $[0,1]$. This is because, for any path $\gamma : [a, b] \to \mathbb{C}$, we can define another path $\gamma^* : [0,1] \to \mathbb{C}$ as follows:

$$\gamma^*(t) = \gamma(a + (b - a)t), \quad t \in [0,1].$$

We can picture γ^* as being the path γ speeded up (or slowed down) so that it runs its course in exactly one second instead of in $b - a$ seconds, and with the stopwatch started at the instant when it departs from its first point. It is clear that the tracks of γ and of γ^* are identical, and that they pass through the same points in the same order (including their shared first point and their shared last point), so they are equivalent in the above sense. We shall soon see (refer to Theorem 6.6.12) that their equivalence runs deeper into the very essence of what we need paths for: because if f is any complex function continuous on the track of γ, then the integrals of f along γ and along γ^* will turn out to be equal. Therefore, for the purposes of complex integration, it makes no difference whether we work with paths having all various domains, or whether we 'normalise them' in this fashion so that $[0, 1]$ becomes the universal standard domain; it is purely a matter of convenience and taste.

6.3.3 Definitions

1. We shall call a path $\gamma : [a, b] \rightarrow \mathbb{C}$ a *smooth path* if it is differentiable on $[a, b]$ and if the derivative γ' is continuous on $[a, b]$. Note (please refer to paragraph 6.2.5 at this point) that since γ is not defined to the left of a, we must take $\gamma'(a)$ as being just the one-sided limit

$$\lim_{h>0, \, h\to 0} \frac{\gamma(a + h) - \gamma(a)}{h}$$

and, for similar reasons, $\gamma'(b)$ as the one-sided limit

$$\lim_{h<0, \, h\to 0} \frac{\gamma(b + h) - \gamma(b)}{h}.$$

2. We shall call a path $\gamma : [a, b] \rightarrow \mathbb{C}$ *piecewise smooth* if the domain $[a, b]$ can be broken up into a finite number of subintervals

$$[a, b] = [a, c_1] \cup [c_1, c_2] \cup [c_2, c_3] \cup \cdots \cup [c_n, b]$$

(overlapping only endpoint-to-endpoint) so that the restriction of γ to each of these subintervals is smooth there.

3. A piecewise smooth path is called a *contour*.

6.3.4 Note
Informally and approximately, what 'smooth' means is that, at each instant of 'time' within the relevant interval, the moving particle that traces out the path is moving in a particular identifiable direction, and that this direction changes continuously as time goes on. What 'piecewise smooth' means is that this continuously changing direction is allowed to break down at a finite number of awkward points along the trajectory, and the geometrical effect of this is to allow a few sharp corners or cusps to occur on the track. The main practical application of this freedom is that it is often convenient to put together several pieces of straight lines and circular arcs in order to build up a useful contour for some

particular integration; the 'piecewise' notion then permits us not to worry about sudden changes of direction occurring where we join two fragments of path, each of which is smooth on its own time interval.

6.3.5 Examples

1. Perhaps the simplest of all paths is a straight line segment. If z_0 and z_1 are any (distinct) complex numbers, then the function $\gamma : [0, 1] \to \mathbb{C}$ given by

$$\gamma(t) = (1 - t)z_0 + tz_1, \quad 0 \le t \le 1$$

is certainly a smooth path (the derivative is the constant $z_1 - z_0$ and is surely continuous) from z_0 to z_1. We shall denote it by $[z_0, z_1]$.

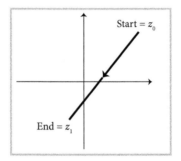

2. Any polygonal figure—a succession of straight segments, each beginning where the previous one ended—can readily be interpreted as a piecewise smooth path in this terminology. For example, the mapping

$$\gamma(t) = \begin{cases} (1 - t)(-2) + t(4 + 4i) & \text{if } 0 \le t \le 1, \\ (2 - t)(4 + 4i) + (t - 1)(-2 + 4i) & \text{if } 1 \le t \le 2, \\ (3 - t)(-2 + 4i) + (t - 2)(1 + i) & \text{if } 2 \le t \le 3, \\ (4 - t)(1 + i) + (t - 3)(-3i) & \text{if } 3 \le t \le 4 \end{cases}$$

is represented by the (open) polygon joining $-2, 4 + 4i, -2 + 4i, 1 + i$ and $-3i$ in that order. We remark that this particular path is neither closed nor simple, and that it is piecewise smooth (with sharp corners at $4 + 4i, -2 + 4i$ and $1 + i$).

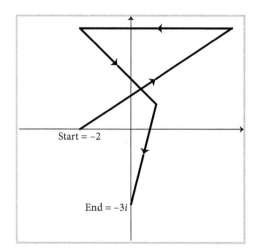

3. A path of particular importance is *the unit circle*: the circle of centre 0 and radius 1, starting and ending at 1, transcribed once anticlockwise. We can define it formally as

$$\gamma(t) = e^{it}, \quad 0 \le t \le 2\pi$$

or, equally well and equivalently,

$$\gamma_1(t) = e^{2\pi i t}, \quad 0 \le t \le 1.$$

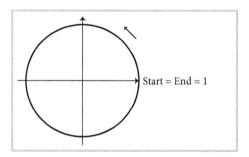

It will be clear already from the picture that it is smooth (not just piecewise smooth) and closed and simple.

4. Any other circle, say, the circle of centre a and radius r, can be given formal specification as a simple, anticlockwise path by slight modifications of the previous item:

$$\beta(t) = a + re^{it}, \quad 0 \le t \le 2\pi$$

or, if preferred,

$$\beta_1(t) = a + re^{2\pi it}, \quad 0 \le t \le 1.$$

Whenever we refer to a circle as constituting a path, the default interpretation (unless explicitly stated otherwise) is that it is to be transcribed once anticlockwise, beginning and ending at the point on its track that has the greatest real part; this style of formal specification is therefore appropriate for it.

5. Take great care to *read the small print* when dealing with circles or parts of circles. If (instead of β as in (4)) we had written $\beta(t) = a + re^{-it}$ $(0 \le t \le 2\pi)$, then the diagram would hardly have changed perceptibly, but the circle would now have been transcribed clockwise, and therefore not have been equivalent to the previous one. Again, if we had written $\beta(t) = a + re^{it}$ $(0 \le t \le 4\pi)$, then the 'same' circle would have been transcribed *twice* anticlockwise, whereas $\beta(t) = a + re^{-it}$ $(0 \le t \le 10\pi)$ gives five clockwise orbits around the unit circle. These are important details and they are easy to misread.

6.4 Combining paths

Draw (on the same piece of paper) a path from some point z_0 to another z_1 and a second path from *that* z_1 to a further z_2, and it is almost difficult to avoid seeing the result as a path from z_0 to z_2.

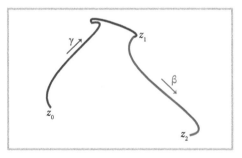

It needs only a little elementary algebra to convert that geometric insight into a proper definition of how to join compatible paths together:

6.4.1 Definition   Let $\gamma : [a, b] \to \mathbb{C}$ and $\beta : [c, d] \to \mathbb{C}$ be any two paths (say, from z_0 to z_1 and from z_1 to z_2 respectively) for which $\gamma(b) = \beta(c)$. We define their *sum* or *concatenation* $\gamma \oplus \beta$ as follows[3]:

---

[3] We have opted for the rather non-standard symbol $\oplus$ here as a warning/reminder that this 'addition' of paths fails to obey many of the standard rules of arithmetic: for instance, there is no expectation that $\gamma \oplus \beta = \beta \oplus \gamma$: if one of those is defined, the other usually is not! Even if both are defined, they are usually not equivalent in the sense of 6.3.2 (2).

$$\gamma \oplus \beta(t) = \begin{cases} \gamma(t) & \text{if } a \le t \le b, \\ \beta(t - b + c) & \text{if } b \le t \le b + d - c. \end{cases}$$

(Notice that the two displayed formulae agree when $t = b$ so that any ambiguity has been prevented: indeed, the 'time' parameter in the second display line has been adjusted in such a way that the β part of the combined path starts immediately the γ component finishes its run, so that β is still allowed a full $d - c$ seconds to complete its trajectory.)

Then $\gamma \oplus \beta$ is a path (from z_0 to z_2) since continuity is not broken at $t = b$. If γ and β happen to be smooth, then $\gamma \oplus \beta$ might not be (because there could be a corner or cusp at $t = b$) but it will be piecewise smooth—indeed, this will be the case even if γ and β themselves are only piecewise smooth.

Returning for a moment to geometric thinking, if we imagine a path from some z_0 to some z_1 and then 'reverse the direction' so that our concept of a moving particle travels through the same points but in the opposite order—

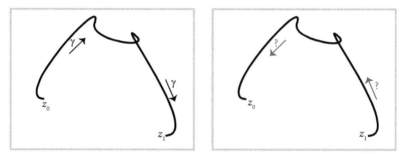

then we appear to create a path from z_1 to z_0. Once again it needs only a tiny adjustment of the algebra to convert the picture into a proper definition:

6.4.2 Definition   For any path $\gamma : [a, b] \to \mathbb{C}$ (say, from z_0 to z_1) the *opposite path* $\ominus\gamma : [-b, -a] \to \mathbb{C}$ is defined by the formula[4]

$$\ominus\gamma(t) = \gamma(-t), \quad -b \le t \le -a.$$

Then $\ominus\gamma$ is a path from z_1 to z_0, and is smooth or piecewise smooth according as γ is.

6.4.3 Remark   If γ and β are two paths from z_0 to z_1, then $\gamma \oplus (\ominus\beta)$ is a *closed* path (starting and ending at z_0). It will be piecewise smooth (and therefore a closed contour) if γ and β are piecewise smooth.

---

[4] The symbol $\ominus$ is non-standard, but matches our choice of $\oplus$ for path sum.

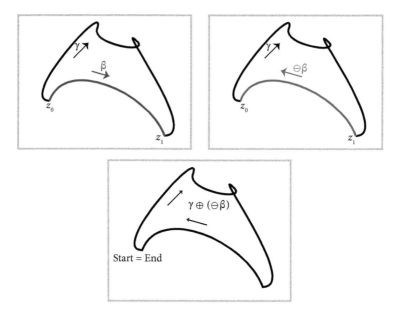

Our primary intention being that of integrating some complex function along a path between two points, it will surely be necessary for such a path to lie within the domain of the function in question, and therefore also for the domain actually to contain paths that join any two points between which we need to carry out an integration exercise. This leads us into a brief discussion concerning the broad 'shape', the *topology* of the kinds of domain that are well suited for these purposes. (Fortunately, all the usual functions do have domains that satisfy these constraints as to overall shape, but we need to be aware of what the constraints are before we can confidently proceed.)

6.5 Connected sets and domains

6.5.1 Definition   A non-empty subset H of $\mathbb{C}$ is called *connected*[5] if each two points in H can be joined by a path lying within H: that is, if for every choice of $x \in H$ and of $y \in H$ there exists a path γ whose first point is x, whose last point is y and whose track is entirely contained in H.

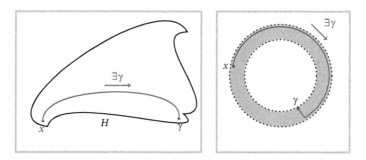

---

[5] More precisely, *path-connected*.

For instance, $C(z_0, r), D(z_0, r)$ and $\overline{D}(z_0, r)$ are (fairly obviously) connected for any choice of z_0 and r, and so is any 'annulus' (that is, the region of the plane lying between two concentric circles).

Simple examples of sets that are *not* connected include the union of two non-overlapping discs, the complement in $\mathbb{C}$ of the real axis, and the set

$$\{z = x + iy \in \mathbb{C} : y = \sin(1/x), x \neq 0\}$$

(the graph of the function $\sin(1/x)$, thought of here as a subset of the complex plane).

A particular class of connected sets that turn up very often in practice are what are called *star-shaped sets*. The term is fairly self-explanatory, but we still need a precise definition for it:

6.5.2 Definition   A non-empty subset S of $\mathbb{C}$ is called *star-shaped* if there exists at least one so-called *star centre* for it: that is, a point z_0 in S with the property that, for every $w \in S$, the straight line segment $[z_0, w]$ that joins z_0 to w lies entirely within S.

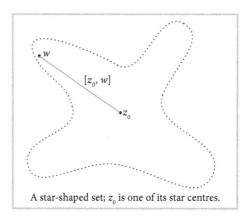

A star-shaped set; z_0 is one of its star centres.

The union of the real axis and the imaginary axis gives a (rather extreme) example of a star-shaped set, one whose *only* star centre is the origin. Any disc (open or closed) is a star-shaped set and, in both cases, any point at all belonging to the disc acts as a star centre for that disc.

6.5.3 Lemma   Every star-shaped set is connected.

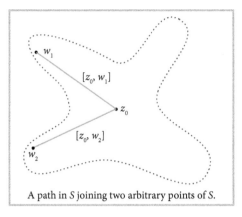

A path in S joining two arbitrary points of S.

A *diagrammatic* proof is almost immediate: if S is star-shaped and z_0 is a star centre for it and w_1 and w_2 are any points of S, then the segments $[z_0, w_1]$ and $[z_0, w_2]$ lie entirely within S, and their union presents us with a path in S joining w_1 and w_2.

The reader who, quite rightly, distrusts the (literally) sketchy nature of that demonstration deserves to be reassured that it can easily be made fully legitimate and based upon our formal definition of the term *path*. Notice first that for any w and z in $\mathbb{C}$ the straight line segment $[w, z]$ joining them, and thought of as being directed from w to z, can be turned into a path $\gamma_{w,z}$ merely by writing

$$\gamma_{w,z}(t) = (1 - t)w + tz, \quad t \in [0, 1]$$

and now the static geometrical object $[w, z]$ is the track of the dynamic path $\gamma_{w,z}$ that corresponds to it. From this point on, whenever we refer to a straight line segment as being a path, this is the path that we have in mind. With this understanding, we see that the path $[w_1, z_0]$ has first point w_1 and last point z_0, that the path $[z_0, w_2]$ has first point z_0 and last point w_2, and therefore that their concatenation $[w_1, z_0] \oplus [z_0, w_2]$ is indeed a path in S from w_1 to w_2. ∎

Quite a bit of time will be saved, especially in more difficult arguments, when we allow ourselves to accept geometric evidence for this kind of path-building, rather than insisting upon formal definitions of the paths being depicted!

The annulus, mentioned earlier, will serve us as a warning that not all connected sets are star-shaped; that is, the converse of the last lemma is false.

6.5.4 Definition   A connected open set (in complex analysis) is usually called a *domain*.

It is arguably a little unfortunate that we now have two different meanings ascribed to the word *domain* but, in practice, this hardly ever causes confusion: for when you identify the domain of any of the complex functions that we wish to work with, it virtually always turns out to be a domain also in the sense defined in this paragraph.

6.6 Integrating along a contour

Recall that a *smooth* path is one which is differentiable and whose derivative is continuous. A *piecewise* smooth path (a contour) is allowed to have finitely many points (corners, cusps...) at which this condition breaks down, and so it can be split at these awkward points into smooth pieces and is, in consequence, the concatenation (the sum) of these smooth 'components'. This possibility of recognising every contour as a sum of smooth paths is the critical link between the two halves of our next definition, and of everything that flows from it.

6.6.1 Definition   Suppose that a complex function f is continuous on a domain G, and that $\gamma : [a, b] \rightarrow G$ is a path in G.

1. If γ is smooth, we define *the integral of f along γ* as

$$\int_\gamma f(z)\, dz = \int_a^b f(\gamma(t))\gamma'(t)\, dt.$$

2. If γ is only *piecewise* smooth, we define *the integral of f along γ* to be the total of its integrals along the smooth components of γ.

6.6.2 Notes

- There is a certain inevitability about this definition that may not be immediately apparent. When we need to integrate $f(z)$ along the path γ, then the only values of f that should be contributing are those that are calculated at points on the track of γ. To make sure of this, we replace 'z' in '$f(z)$' by '$\gamma(t)$': effectively, a change of variable $z = \gamma(t)$. Now that we are looking at $f(\gamma(t))$ and the controlling variable has changed from z to t, the chain rule strongly suggests that we must also replace 'dz' by '$\dfrac{dz}{dt} dt$', that is, by '$\gamma'(t)dt$'. To cover the whole of γ's track, we lastly need to let 'time' t run from a to b, and have now 'discovered' the formula $\int_a^b f(\gamma(t))\gamma'(t)\, dt$.

- Of course this is not a *proof* of the definition—a definition does not need proof in that sense—rather, it will be justified in practice by underpinning a reliable and insightful body of theory, examples and applications (some of which we shall see).

- The appropriateness of our definition of 'smooth' is that, since f and γ are both continuous as is consequently their composite, the continuity of γ' now guarantees that the whole of $f(\gamma(t))\gamma'(t)$ is a continuous function of t, *and therefore can be integrated* (continuous functions always can).

- The natural way to extend the idea from smooth to piecewise smooth is simply to break up piecewise smooth paths at their corners/cusps into smooth fragments, which is exactly what the definition's second part allows.

6.6.3 Example   Integrate the function $f(z) = \dfrac{1}{z-a}$ and the function $g(z) = \dfrac{1}{(z-a)^2}$ around the circle of centre a and radius r, traced once anticlockwise and starting from $a + r$: that is, along the path $\gamma(t) = a + re^{2\pi it}$, $0 \le t \le 1$.

Solution   Since the path is smooth, the definition of the first integral is

$$\int_0^1 f(a + re^{2\pi it})\gamma'(t)\,dt = \int_0^1 \frac{1}{re^{2\pi it}}(2\pi ir)e^{2\pi it}\,dt = \int_0^1 2\pi i\,dt = 2\pi i[t]_0^1 = 2\pi i.$$

The second integral is

$$\int_0^1 g(a + re^{2\pi it})\gamma'(t)\,dt = \int_0^1 \frac{1}{r^2 e^{4\pi it}}(2\pi ir)e^{2\pi it}\,dt = \int_0^1 \frac{2\pi i}{r}e^{-2\pi it}\,dt$$

$$= \frac{1}{r}\left[-e^{-2\pi it}\right]_0^1 = \frac{1}{r}(-1 - (-1)) = 0.$$

6.6.4 Exercise   Verify the claim made in paragraph 6.1 about the integral of $\dfrac{1}{z}$ from $2i$ to $1 - \sqrt{3}i$ along the shorter circular arc.

6.6.5 Example   Let γ be the path sum of the three segments $[0, 1]$, $[1, 1 + i]$ and $[1 + i, i]$ in that order. Integrate z^2 along γ.

Solution

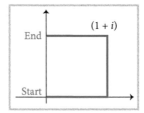

We can easily write down an 'equation of motion' for a moving particle that executes γ, for instance:

$$\gamma(t) = \begin{cases} t & \text{if } 0 \le t \le 1, \\ 1 + (t-1)i & \text{if } 1 \le t \le 2, \\ (3-t) + i & \text{if } 2 \le t \le 3 \end{cases}$$

but notice that γ is only *piecewise* smooth so we have to evaluate the integral along each smooth (straight) component and then total the three partial answers.

1. While $0 \le t \le 1$, we get $\int_0^1 (t)^2 t' \, dt = \int_0^1 t^2(1) \, dt = \left[\frac{1}{3} t^3 \right]_0^1 = \frac{1}{3}$.

2. While $1 \le t \le 2$, we get

$$\int_1^2 (1 + (t-1)i)^2 (1 + (t-1)i)' \, dt = \int_1^2 (2t - t^2 + 2(t-1)i)(i) \, dt$$

$$= \int_1^2 (2 - 2t) + i(2t - t^2) \, dt$$

$$= \left[(2t - t^2) + i\left(t^2 - \frac{1}{3}t^3\right) \right]_1^2 = (0 + \frac{4}{3}i) - (1 + \frac{2}{3}i) = -1 + \frac{2}{3}i.$$

3. While $2 \le t \le 3$, we get

$$\int_2^3 ((3-t) + i)^2 (3 - t + i)' \, dt = \int_2^3 ((8 - 6t + t^2) + i(6 - 2t))(-1) \, dt$$

$$= \int_2^3 -8 + 6t - t^2 + i(2t - 6) \, dt = \left[-8t + 3t^2 - \frac{1}{3}t^3 + i(t^2 - 6t) \right]_2^3$$

$$= -6 - 9i - (-\frac{20}{3} - 8i) = \frac{2}{3} - i.$$

Adding the three contributions gives an overall answer of $-\frac{1}{3}i$.

It is a bit disappointing that evaluating the last integral (of what was a very simple function along a very simple path) took as much labour as it did. What this indicates is that the *definition* of contour integral is better suited to developing general results than it is to the actual calculation of integrals. So let us now make a start on developing some of the promised results and on seeing how they can make integral evaluation easier and quicker.

6.6.6 Theorem   Suppose that F is differentiable on a domain G and that γ is a contour in G from a first point z_0 to a last point z_1. Then

$$\int_\gamma F'(z) \, dz = F(z_1) - F(z_0)$$

(in effect, the 'fundamental theorem of (real) calculus' also works for complex functions).

Proof   Case 1: where γ is smooth.

Considering the path as, say, $\gamma : [a, b] \to G$, we have by definition

$$\int_\gamma F'(z)\, dz = \int_a^b F'(\gamma(t))\gamma'(t)\, dt = \int_a^b \frac{d}{dt} F(\gamma(t))\, dt$$

via the chain rule. Next, processing that as in 6.2.6 and 6.2.4:

$$= \int_a^b \mathrm{Re}\left(\frac{d}{dt}F(\gamma(t))\right) dt + i \int_a^b \mathrm{Im}\left(\frac{d}{dt}F(\gamma(t))\right) dt$$

$$= \int_a^b \frac{d}{dt}\mathrm{Re}\left(F(\gamma(t))\right) dt + i \int_a^b \frac{d}{dt}\mathrm{Im}(F(\gamma(t))\, dt$$

and these *real* integrals can now be simplified by the familiar fundamental-theorem-of-real-calculus idea:

$$= [\mathrm{Re}(F(\gamma(t)))]_a^b + i\,[\mathrm{Im}(F(\gamma(t)))]_a^b$$

$$= \mathrm{Re}(F(\gamma(b))) - \mathrm{Re}(F(\gamma(a))) + i(\mathrm{Im}(F(\gamma(b))) - \mathrm{Im}(F(\gamma(a))))$$

$$= F(\gamma(b)) - F(\gamma(a)) = F(z_1) - F(z_0) \text{ as predicted.}$$

Case 2: where γ is only piecewise smooth.

Suppose now that $\gamma : [a, b] \to G$ suffers failures of smoothness at 'times' $t_1 < t_2 < t_3 < \cdots < t_n$ in (a, b). By using the result of Case 1 over each of its smooth components, we find that $\int_\gamma f(z)\, dz =$

$$F(\gamma(t_1)) - F(\gamma(a)) + F(\gamma(t_2)) - F(\gamma(t_1)) + F(\gamma(t_3)) - F(\gamma(t_2)) + \cdots + F(\gamma(b)) - F(\gamma(t_n))$$

which cancels down to $F(\gamma(b)) - F(\gamma(a)) = F(z_1) - F(z_0)$ once again.   ∎

Let's revisit Example 6.6.5 now that we have better equipment to solve it quickly:

6.6.7 Example   Let γ be the path sum of the three segments $[0, 1]$, $[1, 1 + i]$ and $[1 + i, i]$ in that order. Integrate z^2 along γ.

Solution   We see that γ is a contour in $\mathbb{C}$ from 0 to i and that z^2 is the derivative of $F(z) = \frac{1}{3}z^3$ everywhere in $\mathbb{C}$. By Theorem 6.6.6, the integral equals

$$F(i) - F(0) = \frac{1}{3}i^3 - 0 = -\frac{1}{3}i,$$

in agreement with our earlier (hard-won!) conclusion.

The fact that contour integration is 'linear' (that is, that it respects adding and multiplying by constants) follows routinely from the definition:

6.6.8 Theorem If f and g are continuous on a domain G, γ is a contour in G and K is a constant (real or complex) then

- $\int_\gamma (f(z) + g(z))\, dz = \int_\gamma f(z)\, dz + \int_\gamma g(z)\, dz,$
- $\int_\gamma (Kf(z))\, dz = K \int_\gamma f(z)\, dz.$

Much more interesting is how modifying and combining paths impacts upon path integrals:

6.6.9 Theorem If f is continuous on a domain G and γ is a contour in G, then

$$\int_{\ominus\gamma} f(z)\, dz = -\int_\gamma f(z)\, dz.$$

(You can usefully compare this with the real-integration result $\int_b^a f(x)\, dx = -\int_a^b f(x)\, dx$.)

Proof Case 1: where γ is smooth.

With $\gamma : [a, b] \to G$ and therefore $\ominus\gamma : [-b, -a] \to G$ given by the formula

$$\ominus\gamma(t) = \gamma(-t) \text{ while } -b \le t \le -a,$$

the definition of path integral gives us

$$\int_{\ominus\gamma} f(z)\, dz = \int_{-b}^{-a} f(\ominus\gamma(t))(\ominus\gamma)'(t)\, dt = \int_{-b}^{-a} f(\gamma(-t))(-\gamma'(-t))\, dt$$

(and now substitute $u = -t$)

$$= \int_{u=b}^{u=a} -f(\gamma(u))\gamma'(u)\frac{dt}{du}\, du = \int_b^a f(\gamma(u))\gamma'(u)\, du$$

$$= -\int_a^b f(\gamma(u))\gamma'(u)\, du$$

(because interchanging the limits of integration changes the sign of the real and imaginary parts of the last integral)

$$= -\int_\gamma f(z)\, dz.$$

Case 2: where γ is only piecewise smooth.

The smooth components of $\ominus\gamma$ are the smooth components of γ with their directions reversed (and concatenated in reverse order also). Case 1 showed us that direction reversal changes the sign of the integral along each smooth component, and therefore also of their total: that is, $\int_{\ominus\gamma} f$ differs from $\int_\gamma f$ only by a change of sign. ∎

6.6.10 Theorem   Suppose that f is continuous on a domain G and that γ and β are contours in G such that the last point of γ is the first point of β. Then

$$\int_{\gamma\oplus\beta} f(z)\,dz = \int_\gamma f(z)\,dz + \int_\beta f(z)\,dz.$$

(Compare this with the real-integration rule $\int_a^b f(x)\,dx + \int_b^c f(x)\,dx = \int_a^c f(x)\,dx$.)

Proof   In the case where $\gamma : [a, b] \to G$ and $\beta : [c, d] \to G$ are smooth, but $\gamma \oplus \beta$ has a cusp (corner, etc.) where they join, then $\gamma \oplus \beta : [a, b + d - c] \to G$ is defined by

$$\gamma \oplus \beta(t) = \begin{cases} \gamma(t) & \text{if } a \le t \le b, \\ \beta(t - b + c) & \text{if } b \le t \le b + d - c \end{cases}$$

and the definition requires us to handle the smooth components separately. So

$$\int_{\gamma\oplus\beta} f(z)\,dz = \int_a^b f(\gamma(t))\gamma'(t)\,dt + \int_b^{b+d-c} f(\beta(t - b + c))\beta'(t - b + c)\,dt.$$

Changing the variable in the second of these to $u = t - b + c$, we get

$$\int_a^b f(\gamma(t))\gamma'(t)\,dt + \int_c^d f(\beta(u))\beta'(u)\,du = \int_\gamma f(z)\,dz + \int_\beta f(z)\,dz$$

as desired. Notice in passing that, if there had *not* been a cusp/corner where γ and β met up, this calculation would still have been valid; so it is always legitimate to insert *additional* break points into the description of a contour if it helps us in calculating an integral along it. (See also Exercise 13 at the end of this chapter.)

Routine induction lets us extend this conclusion to the sum of any finite number of smooth paths.

In the general case now, if γ and β are either smooth or piecewise smooth, then the smooth components of $\gamma \oplus \beta$ are[6] those of γ and those of β, and so both

---

[6] This is not quite universally true because if γ and β happen to be travelling in the same direction where they join up, then one of the smooth fragments of $\gamma \oplus \beta$ would consist of the final portion of γ and the initial portion of β together! In this scenario, we can salvage the argument by inserting an additional break point at the junction, as suggested earlier in this proof.

$\int_{\gamma \oplus \beta} f(z)\, dz$ and $\int_{\gamma} f(z)\, dz + \int_{\beta} f(z)\, dz$ are equally formed by adding together the integrals along all these smooth pieces. ∎

6.6.11 Example Using the theorems now becoming available, evaluate the integral of $\dfrac{1}{z}$ along the *other* circular path mentioned in paragraph 6.1—the longer circular arc from $2i$ to $1 - \sqrt{3}i$.

Solution Let γ_1 and γ_2 be the shorter and the longer arcs of the circle $C(0, 2)$ that join $2i$ to $1 - \sqrt{3}i$.

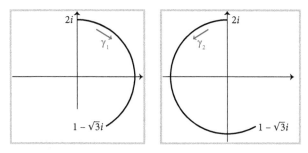

We see that γ_1 is clockwise, γ_2 is anticlockwise, $\ominus\gamma_1$ is anticlockwise and $\gamma_2 \oplus (\ominus\gamma_1)$ is the whole circle $C(0, 2)$ transcribed anticlockwise and starting at $2i$. We hopefully confirmed in 6.6.4 that, for this function, $\int_{\gamma_1}$ is approximately $-2.618i$, and an argument virtually identical to that of the first part of 6.6.3 shows that $\int_{\gamma_2 \oplus (\ominus\gamma_1)} = 2\pi i$. Now use the last two theorems:

$$2\pi i = \int_{\gamma_2 \oplus (\ominus\gamma_1)} = \int_{\gamma_2} + \int_{\ominus\gamma_1} = \int_{\gamma_2} - \int_{\gamma_1} = \int_{\gamma_2} + 2.618i \ approx.$$

from which $\int_{\gamma_2} = 2\pi i - 2.618i = 3.665i$ approximately.

6.6.12 Theorem Suppose that f is continuous on a domain G, that $\gamma : [a, b] \to G$ is a contour in G, and that $\gamma^* : [0, 1] \to G$ is defined by

$$\gamma^*(t) = \gamma(a + (b - a)t), \quad t \in [0, 1]$$

(in less formal language, γ^* is the path γ speeded up or slowed down so as to take just one second). Then

$$\int_{\gamma} f(z)\, dz = \int_{\gamma^*} f(z)\, dz.$$

Proof If γ (and therefore also γ^*) are smooth, then

$$\int_{\gamma^*} f(z)\, dz = \int_0^1 f(\gamma(a + (b - a)t))\gamma'(a + (b - a)t))(b - a)\, dt$$

which, when we substitute $u = a + (b - a)t$, becomes

$$\int_{u=a}^{u=b} f(\gamma(u))\gamma'(u)\frac{du}{dt}\, dt = \int_{u=a}^{u=b} f(\gamma(u))\gamma'(u)\, du = \int_{\gamma} f(z)\, dz$$

as predicted.

When γ and γ^* are only piecewise smooth, this conclusion applies to each smooth component, and the totals give the desired conclusion. ∎

6.6.13 Exercise   If $\gamma : [a, b] \to G$ is a *closed* contour (so $\gamma(a) = \gamma(b)$) and $a < c < b$, we can describe another closed contour $\Gamma : [c, c - a + b] \to G$ as follows:

$$\Gamma(t) = \begin{cases} \gamma(t) & \text{if } c \leq t \leq b, \\ \gamma(t - b + a) & \text{if } b \leq t \leq c - a + b. \end{cases}$$

In effect, this is 'the same contour as γ', except that it starts (and therefore ends) not at $\gamma(a)$, but at $\gamma(c)$ instead. For instance, if γ were the unit circle $C(0, 1)$ transcribed once anticlockwise and starting/ending at 1 ($\gamma(t) = e^{it}$, $0 \leq t \leq 2\pi$) and $c = \pi/2$, then Γ as just described is the unit circle transcribed once anticlockwise but starting/ending at i.

Show that, for any function f continuous on G, $\int_{\gamma} f(z)\, dz = \int_{\Gamma} f(z)\, dz$.

A valuable theorem in real calculus that often gets us out of difficulty when faced with an integral that we cannot see how to do is the one that says: if we are able to spot a constant K such that $|f(x)| \leq K$ for all x such that $a \leq x \leq b$, then $\int_a^b f(x)\, dx$ cannot be greater in modulus than $K(b - a)$. The naturally corresponding complex calculus result says that if $|f| \leq K$ at every point on the track of γ, then $\int_{\gamma} f(z)\, dz$ cannot be greater in modulus than K times the length of γ. To set this up, we first need a proper definition of the length of a contour.

6.6.14 Definition   The *length of a smooth path* $\gamma : [a, b] \to \mathbb{C}$ is defined as

$$L(\gamma) = \int_a^b |\gamma'(t)|\, dt.$$

The *length of a contour* is defined as the total of the lengths of its smooth components.

We can supply some motivation for 6.6.14 by verifying that its definition of length for paths that are built up from parts of straight lines and circles is in agreement with common sense. (In fact, most of the contours we work with in examples in this area are constructed in precisely this way.)

6.6.15 Example As defined by 6.6.14,

1. the length of a segment $[a, b]$ in $\mathbb{C}$, considered as a path, is $|a - b|$;
2. the length of a circular arc of radius r and subtending an angle θ at its centre, considered as a path, is $r\theta$.

Solution

1. To express the segment as a path γ, we can put $\gamma(t) = (1 - t)a + tb$ $(0 \le t \le 1)$. Then $\gamma'(t) = -a + b$ for all relevant t, and 6.6.14 yields

$$L(\gamma) = \int_0^1 |-a + b| \, dt = |a - b|$$

as expected.

2. Consider the path along the arc of the circle $C(a, r)$ that runs anticlockwise from the point $a + re^{i\phi}$ through θ radians. We can express it as $\beta(t) = a + re^{it}$, $\phi \le t \le \phi + \theta$. Since $\beta'(t) = ire^{it}$, 6.6.14 delivers

$$L(\beta) = \int_\phi^{\phi+\theta} |ire^{it}| \, dt = \int_\phi^{\phi+\theta} r \, dt = [rt]_\phi^{\phi+\theta} = r\theta$$

as expected.

6.6.16 Theorem: estimating a contour integral Suppose that f is continuous on a domain that contains the track of a contour γ and that $|f(z)| \le K$, where K is a constant, at every point of the track. Then

$$\left| \int_\gamma f(z) \, dz \right| \le KL(\gamma).$$

Proof (Keep in mind that $\text{Re}(z) \le |z|$ for every complex number z, and that *real integration respects inequalities* in the sense that, if $f(x) \le g(x)$ for all x in $[a, b]$, then $\int_a^b f(x) \, dx \le \int_a^b g(x) \, dx$ also.)

We shall assume that $\gamma : [a, b] \to \mathbb{C}$ is smooth (then the case of the general contour will follow by using the *smooth* special case on each of γ's smooth components).

Express $\int_\gamma f(z)\,dz$ as $re^{i\theta}$: so our task is to estimate the real number r. Now

$$r = e^{-i\theta}\int_\gamma f(z)\,dz = \int_\gamma e^{-i\theta}f(z)\,dz$$

$$= \int_a^b e^{-i\theta}f(\gamma(t))\gamma'(t)\,dt$$

and so

$$r = \mathrm{Re}(r) = \mathrm{Re}\left(\int_a^b e^{-i\theta}f(\gamma(t))\gamma'(t)\,dt\right)$$

$$= \int_a^b \mathrm{Re}\left(e^{-i\theta}f(\gamma(t))\gamma'(t)\right)\,dt \qquad *^7$$

$$\leq \int_a^b \left|e^{-i\theta}f(\gamma(t))\gamma'(t)\right|\,dt$$

$$= \int_a^b |f(\gamma(t))||\gamma'(t)|\,dt \leq \int_a^b K|\gamma'(t)|\,dt = K\int_a^b |\gamma'(t)|\,dt = KL(\gamma). \qquad \blacksquare$$

6.6.17 Remark   Notice that the estimate for $\left|\int_\gamma f\right|$ provided by 6.6.16 is often absurdly over-generous. For instance, if you use it on the integral around $C(0, 10)$ of the function z^{12}, then it guarantees that the modulus of that integral cannot be greater than $20, 000, 000, 000, 000\,\pi$. This is, indeed, absolutely correct; the exact value, however, is 0.

6.6.18 Summary   The official definition of contour integral is expressed alge-braically (by means of the formula $\int_a^b f(\gamma(t))\gamma'(t)\,dt$ for smooth paths, together with the policy of splitting piecewise smooth paths into their smooth pieces and adding up the integrals along these fragments). Yet it is, for most practical pur-poses, determined geometrically also: by the configuration of the track, a clear understanding of where it starts and where it ends, what direction it is travelling in, and (if relevant) 'how many times round the track' we need to go. The reasons underlying this geometrical approach include

- that the integral is independent[8] of the 'speed of travel' (see 6.6.12),

---

[7] See 6.2.6 if in doubt here.
[8] Indeed—although we have not shown this—even variability of speed as we travel along the path does not affect the value of the integral: it is 'independent of parametrization'.

- that in the case of a closed contour, the integral is independent of where we start (see 6.6.13),
- that it is often faster and easier to draw an appropriate contour than it is to write out a full functional specification of it,
- that a geometrical approach may let us easily see paths or path components being added (concatenated) via $\oplus$ or having their directions reversed via $\ominus$,
- that $\int_{\gamma \oplus \beta} = \int_{\gamma} + \int_{\beta}$ (see 6.6.10) and $\int_{\ominus \gamma} = -\int_{\gamma}$ (see 6.6.9),
- that the lengths of most of the contours we actually use can be determined by visual inspection using nothing more sophisticated than elementary school geometry (see 6.6.15).

6.7 Exercises

1. (In connection with 6.2.2) supposing that $\gamma : A \to \mathbb{C}$ is a complex-valued function whose domain A is a set of real numbers and that $p \in A$, prove that $\gamma(t)$ is continuous at p if and only if both $\mathrm{Re}(\gamma(t))$ and $\mathrm{Im}(\gamma(t))$ are continuous at p.

2. (In connection with 6.2.4) supposing that $\gamma : A \to \mathbb{C}$ is a complex-valued function whose domain A is a set of real numbers and that p is an interior point of A, prove that $\gamma(t)$ is differentiable at p if and only if both $\mathrm{Re}(\gamma(t))$ and $\mathrm{Im}(\gamma(t))$ are differentiable at p. Also confirm that, in the case where these conditions hold, $\gamma'(p)$ is $(\mathrm{Re}(\gamma(t)))' + i(\mathrm{Im}(\gamma(t)))'$ evaluated at $t = p$.

3. For each of the paths defined as follows:

$$\gamma_1(t) = \begin{cases} t - i & 0 \le t \le 2, \\ 2 + (t-3)i & 2 \le t \le 4, \\ 6 - t + i & 4 \le y \le 6, \\ (7-t)i & 6 \le t \le 8; \end{cases}$$

$$\gamma_2(t) = \begin{cases} e^{it} & 0 \le t \le \pi, \\ -1 + i + e^{i(t-3\pi/2)} & \pi \le t \le 2\pi; \end{cases}$$

$$\gamma_3(t) = \begin{cases} 1 + 2e^{-it} & -\pi \le t \le 0, \\ 3 - 2it & 0 \le t \le \frac{5}{2}; \end{cases}$$

sketch the curve and determine its length. Also state whether it is smooth, piecewise smooth, closed, simple.

4. Evaluate

$$\int_{-\pi/2}^{\pi/2} (1 + i \sin t)(1 + i \cos t)\, dt.$$

5. Evaluate

$$\int_0^{\pi/2} t^2 e^{it}\, dt.$$

6. Sketch the path γ that is defined as follows:

$$\gamma(t) = \begin{cases} 3i + e^{\pi i(1-t)}, & 0 \le t \le \frac{3}{2}, \\ \left(\frac{7}{2} - t\right) i, & \frac{3}{2} \le t \le \frac{7}{2}. \end{cases}$$

Determine the length of γ and the integral $\int_\gamma z\cos(2z) - z^2 \sin(2z)\, dz$. Also find a constant K (no matter how large) for which you can be certain that

$$\left| \int_\gamma z^6 e^{\cos z}\, dz \right| \le K.$$

(The hyperbolic functions can again be useful in this exercise.)

7. Suppose that $f(z)$ is a continuous complex function defined on a (connected) domain G, and that $g(z) = e^{f(z)}$ is constant everywhere in G. Prove that $f(z)$ also is constant everywhere in G. (Note that the complex exponential function is not one-to-one: for instance, $e^{x+iy} = e^{x+i(y+2\pi)}$ for any values of x and y.)

(*Suggestion:* if f were not constant, we could choose two points z_0 and z_1 in G such that $f(z_0)$ and $f(z_1)$ were unequal. Choose also a path $\gamma : [0,1] \to G$ within G from z_0 to z_1 and consider how $g(z)$ varies along the track of γ.)

8. Sketch the path γ that is defined by

$$\gamma(t) = \begin{cases} e^{-\pi i t}, & 0 \le t \le 1, \\ -1 + i(t-1), & 1 \le t \le 2, \\ t - 3 + i, & 2 \le t \le 3. \end{cases}$$

For this path γ, evaluate
(a) $\int_\gamma (3iz + 7)\, dz$ by the definition of integral along a path,
(b) $\int_\gamma (3iz + 7)\, dz$ using Theorem 6.6.6,
(c) $\int_\gamma (z^2 + \cos z)\, dz$ using Theorem 6.6.6.

9. (a) Find the length of the path γ where $\gamma(t) = (1 + \cos t)e^{it}$, $0 \le t \le \frac{\pi}{2}$.
(b) Estimate the modulus of the integral of z^{10} along γ.

10. Roughly sketch the path defined by

$$\gamma(t) = e^{(1+i)t}, \quad t \in [0, 2\pi]$$

and calculate its length.

11. For the path γ defined by $\gamma(t) = t^2 e^{2\pi i t}$, $1 \le t \le 2$:
    (a) Find the length of γ,
    (b) Noting that $|z| \ge 1$ at every point on the track of γ, obtain a (rough) upper bound for the modulus of the integral $\int_{\gamma} z^{-1} dz$,
    (c) Calculate the exact value of the last-named integral.

12. For real positive constants a and b we consider the path defined by

$$\gamma(t) = a \cos t + ib \sin t, \quad t \in [0, \pi].$$

    (a) Roughly sketch γ in the case where $a = 2$ and $b = 3$, and again in the case where $a = 3$ and $b = 2$.
    (b) Calculate $\int_0^{\pi} \gamma(t) \, dt$.
    (c) Calculate $\int_{\gamma} 1 \, dz$.
    (d) Calculate $\int_{\gamma} z \, dz$.
    (e) Calculate $\int_{\gamma} \bar{z} \, dz$.
    (f) Which (if any) of these calculations could have been done more easily using 6.6.6?

13. Suppose that $\gamma : [a, c] \to \mathbb{C}$ is a path, that $a < b < c$, and that γ is known to be smooth both on $[a, b]$ and on $[b, c]$. We define two paths $\gamma_L : [a, b] \to \mathbb{C}$ and $\gamma_R : [b, c] \to \mathbb{C}$ by the formulae

$$\gamma_L(t) = \gamma(t), \quad a \le t \le b,$$

$$\gamma_R(t) = \gamma(t), \quad b \le t \le c.$$

    Verify that, for any function f that is continuous on the track of γ, we have

$$\int_{\gamma} f(z) \, dz = \int_{\gamma_L} f(z) \, dz + \int_{\gamma_R} f(z) \, dz$$

    whether or not γ is smooth on $[a, c]$.

14. Find the length of the path defined by

$$\gamma(t) = t + \frac{i}{2a} \left(e^{at} + e^{-at} \right), \quad -\lambda \le t \le \lambda$$

    where a and λ are positive constants. (If you wish, you may use the hyperbolic functions here: see Chapter 2, Exercise 13.)
    Calculate, in terms of λ, the value or values of a for which the length of γ turns out to be $\dfrac{2\sqrt{2}}{a}$.

15. For each of the following statements, provide either a proof that it is true, or a counterexample to demonstrate that it is false. (A clear diagram is generally acceptable as a way of presenting a counterexample to such assertions if they actually are false.)

(a) If two star-shaped sets have non-empty intersection, then their intersection is star-shaped.

(b) If two connected sets have non-empty intersection, then their intersection is connected.

(c) If two star-shaped sets have non-empty intersection, then their union is star-shaped.

(d) If two connected sets have non-empty intersection, then their union is connected.

(e) The union of an increasing infinite sequence

$$A_1 \subseteq A_2 \subseteq A_3 \subseteq \cdots \subseteq A_n \subseteq A_{n+1} \subseteq \cdots$$

of connected sets is connected.

(f) [*This one is tricky!*] The union of an increasing infinite sequence

$$A_1 \subseteq A_2 \subseteq A_3 \subseteq \cdots \subseteq A_n \subseteq A_{n+1} \subseteq \cdots$$

of star-shaped sets is star-shaped.

7 Cauchy's theorem(s)

7.1 Introduction

We saw (in 6.6.3) that the integral of z^{-1} around the unit circle was not, despite what a reasonable person might have expected, zero, but rather $2\pi i$. This flies in the face of what we all found to be reliable in real integration, that the integral of any real function from a point *to the same point again* always turned out to be zero. It raises a number of questions, ranging from quite superficial ones (Why is it not zero? Why the particular number $2\pi i$?) to the more general and more reflective (For which functions and which closed contours can we be sure that the integral really is zero? If it's not going to be zero, how can we predict what it will be?). This chapter studies what is called 'Cauchy's theorem', a term which actually refers to a collection of results, each of which says that *in certain circumstances* we can be assured that an integral around a closed contour will be zero. Remember that a function is called *regular* on an open set if it is differentiable at every point of that open set.

7.2 Baby Cauchy

The first such result that we need to take note of (generally not counted as part of Cauchy's theorem since it is so elementary but, nevertheless, one that fits the overall description) is the following:

7.2.1 Theorem Suppose that a function f is regular on an open set that includes (the track of) a closed contour γ. Then

$$\int_\gamma f'(z)\,dz = 0.$$

Proof If $\gamma(a)$ is the first (and therefore also the last) point of γ, then the fundamental theorem of calculus (6.6.6) says that

$$\int_\gamma f'(z)\,dz = f(\gamma(a)) - f(\gamma(a)) = 0. \qquad \blacksquare$$

It will benefit us to spin off a corollary of this right away.

Integration with Complex Numbers. McCluskey and McMaster, Oxford University Press

7.2.2 **Corollary**   Suppose that f is regular on a disc centre a, and that ε is a given positive number. Then we can find a (smaller) disc D centred on a such that, for any triangular[1] contour γ lying within D and having a either on its track or in its interior, we get[2]

$$\left| \int_\gamma f(z)\,dz \right| \le \frac{1}{2}\varepsilon(L(\gamma))^2.$$

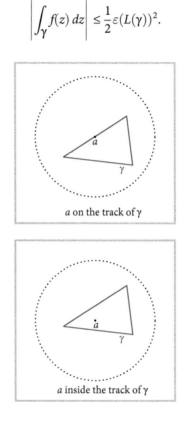

a on the track of γ

a inside the track of γ

Proof   Since f is differentiable at a, we know that

$$\frac{f(z) - f(a)}{z - a} - f'(a) \to 0 \text{ as } z \to a$$

so, putting $g(z) = \dfrac{f(z) - f(a)}{z - a} - f'(a)$ (for $z \ne a$) together with $g(a) = 0$, we see that g is continuous and will be smaller in modulus than ε provided only that z is close enough to a: say, $|z - a| < \delta$ for some choice of $\delta > 0$. Throughout the disc

---

[1] Essentially the same argument works for any closed contour, as soon as we properly define the idea of a point being *inside* such a contour. See, for example, paragraph 7.5.1.

[2] Recall that $L(\gamma)$ denotes the length of γ.

D of centre a and radius δ we then have

$$f(z) = f(a) + (z - a)f'(a) + (z - a)g(z)$$

and, for a triangular contour γ as described,

$$\int_\gamma f(z)\, dz = \int_\gamma f(a)\, dz + \int_\gamma (z - a)f'(a)\, dz + \int_\gamma (z - a)g(z)\, dz.$$

Now on the right-hand side, the first and second integrals both equal zero by 7.2.1 because their integrands are derivatives (of pretty obvious functions). Consider the third: while z traverses the track of γ, the distance $|z - a|$ from a to z has to be less than half the length of γ, and $g(z)$ has to be smaller in modulus than ε,

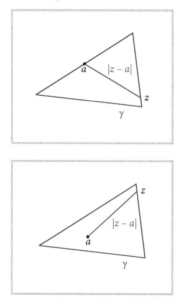

so, now invoking the 'estimation of integral' result from 6.6.16, the third integral is smaller in modulus than $\frac{1}{2}L(\gamma) \cdot \varepsilon \cdot L(\gamma) = \frac{1}{2}\varepsilon(L(\gamma))^2$ as asserted.   ∎

7.2.3 Note   Since their importance is perhaps less than totally transparent, it may be worthwhile to reformulate the last two results in (or closer to) plain English:

1. The integral of any derivative around a closed contour will always be exactly zero.

2. The integral of a regular function around a triangular contour can be made smaller (in modulus) than any chosen multiple of the square of the path length, just by forcing the contour to wrap sufficiently closely around a point where differentiability holds.

3. Keep in mind that where we're going with this is a guarantee that the integral of a 'regular' (in a sense that we haven't yet made quite precise enough) function around a closed contour will be exactly zero: so 1 and 2 are steps in the right direction.

7.3 **The triangular contour case**

Here is the next step towards a versatile version of Cauchy's theorem:

7.3.1 Cauchy's theorem—the triangular contour case   Suppose that f is regular on an open set that contains a triangular contour Δ and its interior. Then $\int_\Delta f(z)\, dz = 0$.

Proof   (This is not a difficult proof but it is rather lengthy, so here's a roadmap of how we are going to proceed. Broadly speaking, our strategy is the following:

1. We shall *assume* that the integral is non-zero.

2. We successively divide Δ into smaller and smaller sub-triangles, each of which carries a large share of that integral—so the integral around the typical sub-triangle is *relatively big*.[3]

3. We eventually squeeze one of these sub-triangles inside a disc so small that we are able to use 7.2.2 on it—so the integral around that triangle is *relatively very small*.

4. The *contradiction* now arising from steps 2 and 3 shows that our non-zero assumption in step 1 must have been wrong—hence the result.

Now let us implement that strategy.)

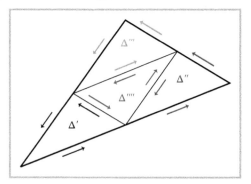

Suppose that $I = \left| \int_\Delta f(z)\, dz \right|$ is non-zero (and therefore strictly positive).

Imagine that we join (by straight line segments) the midpoints of the three sides of Δ, carving it up into four half-sized sub-triangles $\Delta', \Delta'', \Delta'''$ and Δ''''. If we

---

[3] Of course, any reference to 'big' or 'small' in the complex setting always refers to the *modulus* of the quantity, since the complex numbers themselves are not ordered in any meaningful way.

integrate f around these four smaller triangular closed contours and add, notice that the back-and-forth integrals along our construction lines cancel out, so the total is just the original integral $\int_\Delta f$. Therefore at least one of these sub-triangular integrals must contribute a quarter or more of that total. In other words, we can pick one of the four half-sized triangles (let us call it Δ_1) for which

$$\left| \int_{\Delta_1} f(z)\, dz \right| \geq \frac{1}{4} I.$$

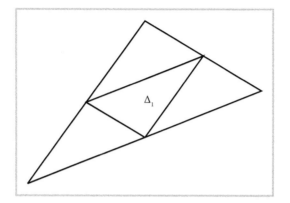

Next, we do exactly the same on Δ_1: we carve it up into four quarter-sized triangles (by joining the midpoints of its sides) and argue that at least one of these (let's call it Δ_2) must contribute a quarter or more to the integral around Δ_1. That is,

$$\left| \int_{\Delta_2} f(z)\, dz \right| \geq \left(\frac{1}{4} \right)^2 I.$$

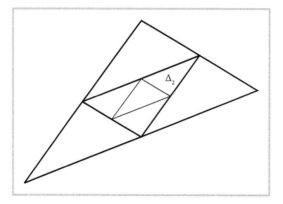

We now find ourselves in a mindlessly repetitive process that, thanks to induction, generates an unending sequence of smaller and smaller triangles

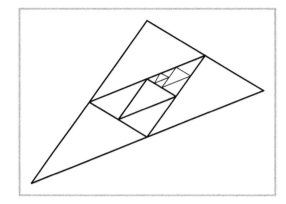

$$\Delta, \Delta_1, \Delta_2, \Delta_3, \cdots \Delta_n, \cdots,$$

obeying, for each positive integer n, the inequality

$$\left| \int_{\Delta_n} f(z)\, dz \right| \geq \left(\frac{1}{4} \right)^n I.$$

Since these triangles constitute a nest of closed sets (see paragraph 4.2.19) there is a point a that lies in or on Δ_n for every n. This will be a point at which f is regular so, using 7.2.2 with any chosen value of ε, there is a disc D enjoying the property set out in 7.2.2. Yet since every Δ_n includes a on its track or in its interior, and the 'size' (whether we assess it by length or diameter) of Δ_n is decreasing by half each time we subdivide, *ergo* the size is tending to zero, and a stage will come at which Δ_n lies entirely inside the disc D.

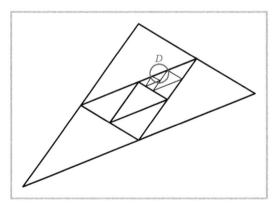

Hence

$$\left| \int_{\Delta_n} f(z)\, dz \right| \leq \frac{1}{2} \varepsilon (L(\Delta_n))^2.$$

Of course the length of Δ_n is just $L(\Delta)/2^n$, so the last display can be rewritten as

$$\left| \int_{\Delta_n} f(z)\,dz \right| \le \frac{1}{2}\varepsilon \left(\frac{1}{4} \right)^n (L(\Delta))^2.$$

Lastly, combining this with a previous display, we see that

$$I \le \frac{1}{2}\varepsilon(L(\Delta))^2$$

for *every choice* of the positive number ε. This is impossible since I was assumed to be a strictly positive number. Hence it cannot have been, and the integral around Δ can only have been zero. ■

7.4 The star domain case

At the risk of stating the obvious, not all contours are triangular, so we need to go a lot more general than 7.3.1. This is a good moment at which to recall (from 6.5.2) what a *star-shaped* domain (or *star domain*) is: it's an open set G for which we can find a *star centre*, that is, a point a such that, for every $z \in G$, the straight line segment $[a, z]$ lies entirely inside G.

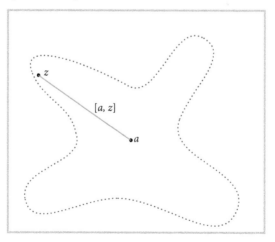

Let us also recall the key step in the proof of the 'real' version of the fundamental theorem of calculus. Given a continuous function f defined on a bounded closed interval $[a, b]$, we have to invent another function F whose derivative coincides with f at every point of (a, b) (and then we go on to show that the integral of f is just $F(b) - F(a)$, of course). We are now about to import this argument into the complex plane (*mutatis mutandis!*) to see what good it does us. The main changes are (i) that we operate in a star domain instead of in an interval, and (ii) we assume regularity at the start instead of just continuity.

Here is what we find:

7.4.1 Lemma   Suppose that f is a regular function on a star domain G. Then there is another function F (regular on G) such that $F' = f$ everywhere in G.

Proof   Choose a star centre a for G. For each $z \in G$ note that the segment $[a, z]$ $\subseteq G$ so it is legitimate to define

$$F(z) = \int_{[a,z]} f(w)\, dw$$

(where the segment is now considered as a path from a to z). We focus on a particular (but arbitrary) element z_0 of G and, G being open, we can find an open ball $B(z_0, \delta)$ of sufficiently small radius that it lies entirely within G. For any complex number h of (non-zero) modulus smaller than δ, the triangle joining z_0 and $z_0 + h$ and a lies entirely inside G, so we can apply 7.3.1 to it, deducing that the integral of f around (all three sides of) the triangle is zero.

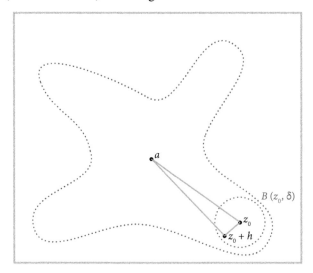

We thus learn (keeping in mind that the two 'long' sides of this triangle are described in opposing directions, one away from a and one towards a, and that reversing the direction of a path changes the sign of the integral along it, as verified in 6.6.9) that

$$0 = \int_{[a,z_0]} f + \int_{[z_0,z_0+h]} f + \int_{[z_0+h,a]} f = \int_{[a,z_0]} f + \int_{[z_0,z_0+h]} f - \int_{[a,z_0+h]} f$$

$$= F(z_0) + \int_{[z_0, z_0+h]} f - F(z_0 + h)$$

and therefore that *the thing whose limit we need to find in order to differentiate*:

$$\frac{F(z_0 + h) - F(z_0)}{h}$$

is equal to

$$\frac{\int_{[z_0, z_0+h]} f}{h}.$$

Since we expect that the limit ought to be $f(z_0)$, we look at the difference

$$\frac{F(z_0 + h) - F(z_0)}{h} - f(z_0) = \frac{\int_{[z_0, z_0+h]} f(w)\, dw}{h} - f(z_0)$$

$$= \frac{\int_{[z_0, z_0+h]} f(w)\, dw}{h} - \frac{\int_{[z_0, z_0+h]} f(z_0)\, dw}{h}$$

(... now there is a line that often confuses the learner. The 'dw's are reminding us that these are integrals along the segment $[z_0, z_0 + h]$ and that w is our name for the variable that transcribes that segment and that $f(w)$ is therefore variable. But z_0 itself is (temporarily) fixed, so $f(z_0)$ is just a constant and, when you integrate any constant K along a segment such as the one we are handling, all you get is Kh. That's why the previous line is *true*. For the reason why the previous line is *useful*, read on...)

$$= \frac{\int_{[z_0, z_0+h]} (f(w) - f(z_0))\, dw}{h},$$

and therefore

$$\left| \frac{F(z_0 + h) - F(z_0)}{h} - f(z_0) \right| = \frac{\left| \int_{[z_0, z_0+h]} (f(w) - f(z_0))\, dw \right|}{|h|}$$

$$\leq \frac{(\max_{[z_0, z_0+h]} |f(w) - f(z_0)|)\, |h|}{|h|} = \max_{[z_0, z_0+h]} |f(w) - f(z_0)|$$

(using 6.6.16) which tends to zero as $h \to 0$ merely because f is continuous at z_0. (Continuity is also why this expression must actually have a maximum value on this closed, bounded segment.) In other terms,

$$F'(z_0) = \lim_{h \to 0} \frac{F(z_0 + h) - F(z_0)}{h} = f(z_0)$$

as claimed. ∎

The next step is the (hopefully) obvious one:

7.4.2 Cauchy's theorem—star domain case   If f is regular on a star domain G and γ is any closed contour lying within G, then

$$\int_{\gamma} f(z)\, dz = 0.$$

Proof   By 7.4.1, f is the derivative of a regular function, so now the result is immediate from 7.2.1.   ∎

7.4.3 Examples

1. Show that if f is regular on an open disc $D(a, r)$ and γ is any closed contour in $D(a, r)$ then $\int_{\gamma} f(z)\, dz$ must be zero.

2. Recall that the technical term *cut plane* means the whole complex plane except for the non-positive real numbers. (The mental image is that someone has taken a very sharp pair of scissors and cut $\mathbb{C}$ along the real interval $(-\infty, 0]$!)

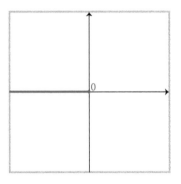

If f is any function regular on the cut plane and γ is any closed contour in the cut plane, show that $\int_{\gamma} f(z)\, dz$ must be zero.

3. Let γ be the (anti-clockwise) polygonal path indicated in the diagram:

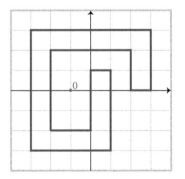

Prove that $\int_\gamma \frac{1}{z}\, dz = 0$.

Solutions

1. The only thing we need to notice is that $D(a, r)$ is a star domain. Then the result is immediate from 7.4.2.

2. Although $\mathbb{C} \setminus (-\infty, 0]$ is not what most people regard as being star-shaped, nevertheless it is a star domain: for instance, the straight line segment from 1 to any point in the cut plane lies entirely within the cut plane, so 1 acts as a star centre for it. Now the result is immediate from 7.4.2.

3. This is (apparently) more awkward since 0 is a point at which $\frac{1}{z}$ fails to make sense, and there is no star domain containing the track of γ but not including 0. However, with a little geometrical insight we can break the question down into pieces that do fit inside easily chosen star domains. For instance, let γ_1, γ_2 and γ_3 be as indicated (in blue and red) in the following diagrams:

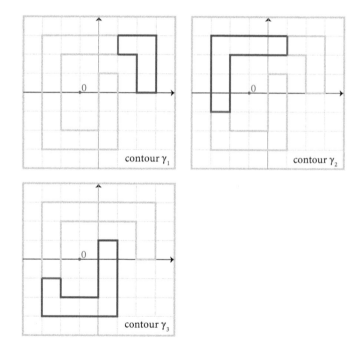

contour γ_1 contour γ_2

contour γ_3

By inserting the extra 'construction lines' (the red lines) we have made from γ these three slightly simpler closed paths represented here. If γ_1, γ_2 and γ_3 are each transcribed anticlockwise then we transcribe the whole of γ anticlockwise but, in addition, each (red) construction line is transcribed twice but in opposite

directions (so the integrals along the construction lines cancel one another out).
Therefore we find that

$$\int_{\gamma_1} f(z)\,dz + \int_{\gamma_2} f(z)\,dz + \int_{\gamma_3} f(z)\,dz = \int_{\gamma} f(z)\,dz.$$

Lastly, it is easy to find three star-shaped domains (for instance, the rectangular
and L-shaped regions indicated in green in the diagrams) each containing one
of the γ_i but excluding 0:

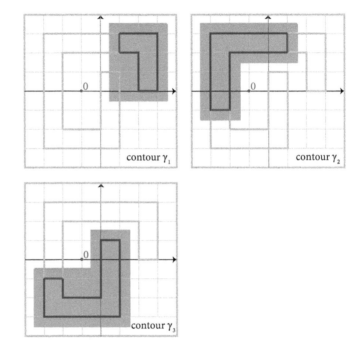

contour γ_1

contour γ_2

contour γ_3

Now 7.4.2 gives $\int_{\gamma_i} f(z)\,dz = 0$ for each i, and the desired conclusion $\int_{\gamma} f(z)\,dz = 0$ follows.

7.5 The general case

Although the techniques established and tried out in Section 7.4 will probably deal
with any specific example of Cauchy's theorem in action that you are ever likely to
come across, there is a more general version that you ought to know about. In its
favour are:

1. that it is the version of Cauchy that is generally used in developing other theorems and other theory,

2. that it is often easier to apply than Section 7.4's technology.

In its disfavour is that it is a lot harder to prove than what we have dealt with in Sections 7.2, 7.3 and (especially) 7.4. To try to extract the best from both worlds, we propose

1. to explain it and (where appropriate) use it, but
2. to make no attempt to present a proper proof.

7.5.1 The Jordan curve theorem For every simple closed contour $\gamma : [a, b] \to \mathbb{C}$, the complement of the track of γ, $\mathbb{C}\backslash\gamma([a, b])$, consists of two disjoint connected open sets, one of which (called the *interior* of γ) is bounded, and the other of which (called the *exterior* of γ) is unbounded. In slightly different language, every simple closed contour cuts the complex plane into two disjoint connected open sets.

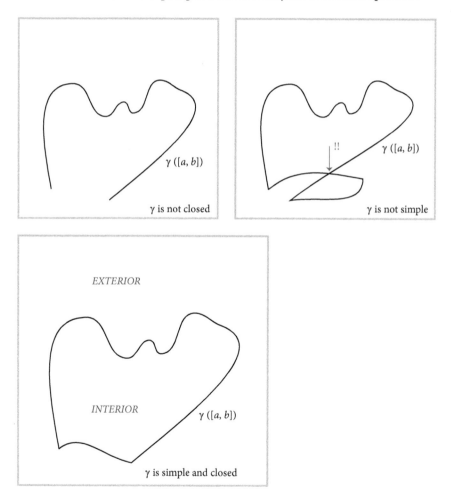

Sketching a handful of simple closed contours tends to make this result look totally obvious and therefore (?) easy to prove. It seriously isn't either.

Notation   The phrase 'f is regular in and on γ' (where γ is some simple closed contour) means that f is regular on some open set that contains both the track of γ and its interior.

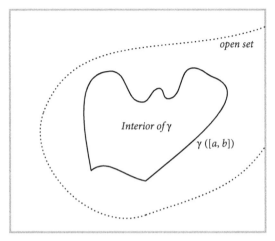

7.5.2 **Cauchy's theorem–general case**   If a function f is regular in and on a simple closed contour γ, then

$$\int_\gamma f(z)\, dz = 0.$$

7.5.3 **Note**   Observe how 7.5.2 deals with the uses of Cauchy's theorem that we encountered previously:

- The triangular contour result 7.3.1 is the special case of 7.5.2 in which we take γ as the triangular contour that we denoted by Δ in 7.3.1.
- The star domain result 7.4.2, in the case where the contour called γ in that paragraph is simple, is immediate from 7.5.2.
- The third example in paragraph 7.4.3 does not need to be geometrically carved up if we use 7.5.2 instead of the star domain version: for it is immediate that the polygonal path involved is a simple closed contour not having 0 in its interior, and the integral is almost immediately seen to be zero: for the domain of $1/z$ – the whole complex plane except for the point 0 – is an open set containing the contour and its interior.
- Notice also how the general version of Cauchy's theorem pays dividends in proofs such as that of 7.6.1 in the next section.

7.6 Cauchy's integral formula

By this stage in the narrative we have a reasonably comprehensive answer to the first of the general questions posed at the beginning of the chapter—'For which functions and which closed contours can we be sure that the integral (of the function around the contour) really is zero?', and so it is time to begin considering the second one—'If it's not going to be zero, how can we predict what it will be?'. For γ a simple closed contour, Cauchy's theorem tells us that $\int_\gamma f(z)\,dz$ will equal zero if f is regular everywhere in and on γ, so regularity has to be broken somewhere if we now want to explore cases in which the integral is non-zero. The easiest, most elementary way to break regularity is to replace $f(z)$ by $\dfrac{f(z)}{z-a}$ where a is a single point in the interior of γ. It turns out that there is a simple and useful answer to what the value of the integral now becomes, and that we can use Cauchy's theorem to establish the result. We shall conclude Chapter 7 by investigating this; for a fuller answer to the second question going well beyond this special case, please see Chapter 9.

7.6.1 Lemma Suppose that γ is a simple closed anticlockwise contour and that the function F is regular in and on γ *except* at a single point a in the interior of γ. Also consider a circle C centred on a whose radius δ is small enough to ensure that C lies entirely within the interior of γ. Then

$$\int_\gamma F(z)\,dz = \int_C F(z)\,dz$$

where C is transcribed once anticlockwise.[4]

Proof Imagine joining two points on C to two points on the track of γ by straight lines as suggested by the following diagram.

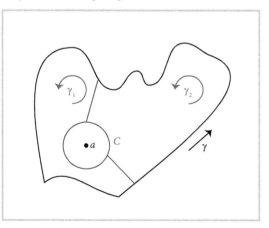

---

[4] In accordance with the default convention—we recall paragraph 6.3.5 (4).

The region enclosed by γ is thus divided up into three, whose (anticlockwise) bounding curves we shall call γ_1, γ_2 and, of course, C itself. Since a is the only point throughout this region at which $F(z)$ fails to be regular, Cauchy's theorem applies to this function with respect to γ_1 and γ_2. Hence we know that

$$\int_{\gamma_1} F(z)\, dz = 0 \quad \text{and} \quad \int_{\gamma_2} F(z)\, dz = 0.$$

When we add these together, notice that the whole of γ will be integrated around anticlockwise, but that each of the joining straight lines is integrated along *twice and in opposite directions*, while C is integrated around *clockwise*. Bearing in mind that a reversal of the direction of integration changes the sign of the answer, we therefore get

$$\int_{\gamma} F(z)\, dz - \int_{C} F(z)\, dz = 0,$$

in other words, that[5]

$$\int_{\gamma} F(z)\, dz = \int_{C} F(z)\, dz.$$

∎

7.6.2 **Cauchy's integral formula**   Suppose that γ is a simple closed anticlockwise contour, that the function f is regular in and on γ and that a is a point in the interior of γ. Then

$$\int_{\gamma} \frac{f(z)}{z-a}\, dz = 2\pi i\, f(a).$$

Proof   Take a circle C as in 7.6.1, in which we saw that $\int_{\gamma}$ was equal to $\int_{C}$. So now we need to calculate $\int_{C}$.

Using the known fact (6.6.3) that $\int_{C}(z-a)^{-1}\, dz = 2\pi i$, we have

$$\int_{C} \frac{f(z)}{z-a}\, dz - 2\pi i\, f(a) = \int_{C} \frac{f(z)}{z-a}\, dz - f(a)\int_{C} \frac{1}{z-a}\, dz$$

$$= \int_{C} \frac{f(z)-f(a)}{z-a}\, dz.$$

Since the last integrand has a limiting value of $f'(a)$ as z approaches a, we can force it as close to $f'(a)$ as we please just by taking the radius δ of C small enough. For instance, we can ensure that

---

[5] Look at the power of the general version of Cauchy's theorem! We've just swopped an integral around some unknowable random contour for an integral around a tiny controllable circle. This is the sort of thing we had in mind when we said that the general Cauchy theorem is useful for developing more theorems.

$$\left| \frac{f(z) - f(a)}{z - a} \right| \leq 1 + |f'(a)|$$

throughout C provided that $\delta \leq$ (some suitably chosen) δ_1. This in turn will guarantee via 6.6.16 that

$$\left| \int_C \frac{f(z) - f(a)}{z - a} \, dz \right| \leq (1 + |f'(a)|)L(C) = 2\pi\delta(1 + |f'(a)|),$$

and therefore that

$$\left| \int_\gamma \frac{f(z)}{z - a} \, dz - 2\pi i f(a) \right| \leq 2\pi\delta(1 + |f'(a)|)$$

for any sufficiently small positive δ. Since the left-hand side is independent of δ (which we can allow to tend to zero) this is only possible if the left-hand side is exactly zero. Hence the result. ∎

7.6.3 Notes

1. Of course, if the contour in Cauchy's integral formula is traversed clockwise instead of anticlockwise, the change of sign gives an integral of $-2\pi i f(a)$ instead. (This was a point that we had no need to labour in discussing Cauchy's theorem itself, since minus zero is still zero.)

2. What the formula also shows us is that $f(a)$ can be evaluated if we can carry out that integration around the contour γ. So the values of a regular function are completely determined within a region if we know what they are around its (simple, closed) boundary!

7.6.4 Examples   Evaluate the following integrals, using Cauchy's integral formula and interpreting each contour as simple and anticlockwise.

1.

$$\int_{|z-1|=2} \frac{\sin z}{2z - \pi} \, dz,$$

2.

$$\int_{|z-2i|=2} \frac{z^3}{z^2 + 9} \, dz,$$

3.

$$\int_{|z+3|=4} \frac{\sin(\pi z)e^{iz}}{(z + 1)^2} \, dz.$$

Solutions

1. The singularity at $\pi/2$ lies within the circle $|z-1| = 2$ so this is an application of Cauchy's integral formula with $f(z) = \sin(z)/2$ and $a = \pi/2$. Thus, the integral evaluates to $2\pi i f(\pi/2) = 2\pi i(\frac{1}{2}) = \pi i$.

2. Since $z^2 + 9$ factorises as $(z + 3i)(z - 3i)$, and $3i$ lies within the circle $|z - 2i| = 2$ but $-3i$ does not, we can rewrite this as

$$\int_{|z-2i|=2} \frac{z^3/(z + 3i)}{z - 3i}\, dz,$$

and perceive it as an application of the formula with $f(z) = z^3/(z + 3i)$ (and $a = 3i$). Hence the integral equals $2\pi i f(3i) = 2\pi i(-27i/(6i)) = -9\pi i$.

3. This does not, at first sight, appear to be a Cauchy-amenable problem since the term on the bottom line is squared. However, look first at the function $g(z) = \dfrac{\sin(\pi z)}{z + 1}$. An easy application of l'Hôpital's rule[6] shows this to have a limit of $-\pi$ as z approaches -1. If we therefore define a new function G to coincide with g everywhere except at -1, and to take the value $-\pi$ at -1, we see that G is continuous at -1 and, indeed, everywhere else. It will be seen later[7] that a function 'patched' in this way is not only continuous there, but regular as well. Now we can rewrite the problem as

$$\int_{|z+3|=4} \frac{G(z)e^{iz}}{z + 1}\, dz$$

and immediately evaluate it (via Cauchy's integral formula, and noting in passing that -1 lies interior to the relevant circle) as $2\pi i G(-1)e^{-i} = 2\pi i(-\pi)e^{-i} = -2\pi^2(\sin 1 + i\cos 1)$.

Lemma 7.6.1 is one of a range of results saying, in one form or another, that the integral of a function along a simple closed contour has some measure of independence from the contour itself! Here, for future use, is another of these:

7.6.5 Lemma   Suppose that f is regular throughout a domain G except for one single point a of G and that γ and β are two simple closed contours in G each of which has a in its interior. Then

$$\int_{\gamma} f(z)\, dz = \int_{\beta} f(z)\, dz.$$

---

[6] See paragraph 11.2.1 as to why this well-known rule from 'real' calculus is also valid for complex functions.

[7] There is a more thorough discussion of this technique in Chapter 9; see, in particular, Theorem 9.1.6.

Proof   Take a circle C centred on a that is small enough to fit into both the interior of γ and the interior of β. Then 7.6.1 shows that $\int_\gamma f(z)\,dz = \int_C f(z)\,dz = \int_\beta f(z)\,dz$. ∎

7.7 Exercises

Throughout this set of exercises, a symbol of the form $C(z_0, r)$ represents the circle of centre z_0 and radius r, as a simple closed contour of integration transcribed once anticlockwise, starting and ending at the point $z_0 + r$.

1. Suppose that γ_1 and γ_2 are two paths with the same first point z_0 and the same last point z_1, and that we can find a star domain G that contains both of their tracks and on which a function f is regular. Prove that

$$\int_{\gamma_1} f(z)\,dz = \int_{\gamma_2} f(z)\,dz.$$

2. (a) Carefully sketch the path γ defined by $\gamma(t) = t + (1 - t^3)i$, $0 \le t \le 1$ and the quarter-circular path $\delta(t) = e^{\pi i(1/2-t)}$, $0 \le t \le 1/2$.

   (b) Write out and simplify the definition of $\int_\gamma z^{-1}\,dz$.

   (c) Identify a star domain containing the tracks of both γ and δ and on which z^{-1} is regular.

   (d) Hence evaluate

$$\int_0^1 \frac{1 + 2t^3}{1 + t^2 - 2t^3 + t^6}\,dt.$$

3. Evaluate the following integrals:

   (a)

$$\int_{C(-3,3)} \frac{\sin(\pi z/8)}{z^3 - 2z + 4}\,dz,$$

   (b)

$$\int_{C(3i,3)} \frac{\sin(\pi z/8)}{z^3 - 2z + 4}\,dz,$$

   (c)

$$\int_\gamma \frac{\sin(\pi z/8)}{z^3 - 2z + 4}\,dz$$

   where γ (anticlockwise) is composed of the lower half of the circle $C(3, 3)$ and its bounding diameter $[0, 6]$. You need not simplify the answers to parts (b) and (c).

4. Evaluate the integrals

$$\int_{C(-2i,1)} \frac{z^3}{z+4i}\, dz \text{ and } \int_{C(-2i,3)} \frac{z^3}{z+4i}\, dz.$$

5. Evaluate the integrals

$$\int_{C(3i,2)} \frac{e^z}{\sin z}\, dz \text{ and } \int_{C(2,1)} \frac{e^z}{(z-\pi/2)\sin z}\, dz.$$

6. Evaluate the integrals

$$\int_{C(2i,2)} \frac{\cos(\pi i z/4)}{z^2+1}\, dz \text{ and } \int_{C(0,1)} \frac{ze^z}{z^3 - 2z^2 + 9z - 18}\, dz.$$

7. (a) Carefully sketch the paths γ and α defined by

$$\gamma(t) = t + i(1-t^2), \quad -1 \le t \le 1 \text{ and } \alpha(t) = e^{-\pi i t}, \quad 1 \le t \le 2.$$

(b) Verify that

$$\int_\gamma \frac{1}{z}\, dz = \int_{-1}^1 \frac{(2t^3 - t) - i(t^2 + 1)}{t^4 - t^2 + 1}\, dt.$$

(c) Find a star domain that contains both their tracks and on which $f(z) = \dfrac{1}{z}$ is regular. (A clear diagram will be sufficient here.)

(d) Hence evaluate

$$\int_{-1}^1 \frac{t^2 + 1}{t^4 - t^2 + 1}\, dt.$$

8. Use Cauchy's integral formula to evaluate:

(a)

$$\int_{C(0,3)} \frac{(z+i)\cos(\pi z)}{z}\, dz,$$

(b)

$$\int_{C(-i,1)} \frac{z^4 + 3iz^2 - z + 2i + 1}{4z + i}\, dz,$$

(c)

$$\int_{C(-2-i,2)} \frac{e^{\sin z}}{z^2 - 9}\, dz.$$

9. Using the technical trick shown in 7.6.4 (3) if and when necessary, evaluate

(a)

$$\int_{C(1,1)} \frac{\sin\left(\frac{\pi z}{3}\right)}{16z^4 - 81} \, dz,$$

(b)

$$\int_{C(1,1)} \frac{\cos\left(\frac{\pi z}{3}\right)}{(16z^4 - 81)^2} \, dz.$$

10. Use Cauchy's integral formula to evaluate

$$\int_{C(4,4)} \frac{1 - \cos(2\pi z)}{(z^3 - 1)^3} \, dz.$$

(*Hint:* reconsider the argument of Example 7.6.4 (3).)

8 Taylor's theorem

8.1 Introduction

Recall once more that a function differentiable at every point of an open set G in the complex plane is said to be *regular on G*. The central insight of this chapter is that, for an open disc D, the concepts 'function regular on D' and 'power series convergent absolutely at each point of D' are virtually interchangeable: if you have one of these, you are guaranteed to have the other also. Much of the usefulness of this is due to the fact that we can carry out term-by-term differentiation on any power series. To revise briefly:

- A complex power series $\sum c_n(z-a)^n$ will generally have a *radius of convergence* (let us denote it by r) such that the series converges absolutely at every point interior to its *circle of convergence* $C(a, r)$ but diverges at every point exterior to that circle. The exceptional types are (i) certain such series converge *only* at $z = a$ and are said to have radius of convergence 0, while (ii) certain important such series converge absolutely at *every* point of the complex plane, and for these the radius of convergence is said to be infinite.

- *Term-by-term differentiation* works for complex power series just as it does for real series: that is, if $\sum c_n(z-a)^n$ has non-zero radius of convergence r and $f(z)$ denotes its sum at the typical point z interior to $C(a, r)$, then f is differentiable there, and $f'(z) = \sum nc_n(z-a)^{n-1}$. So:

$$\text{if } f(z) = c_0 + c_1(z-a) + c_2(z-a)^2 + c_3(z-a)^3 + c_4(z-a)^4 + \cdots$$

$$\text{then } f'(z) = \qquad c_1 + 2c_2(z-a) + 3c_3(z-a)^2 + 4c_4(z-a)^3 + \cdots.$$

We have mentioned from time to time the possibility, and the desirability, of representing functions by sums of power series and of Taylor's theorem as a tool for doing precisely that, so we had better devote a page to revising in some detail what Taylor says in the real scenario, which is as follows:

8.1.1 Taylor's theorem (real case)
Suppose that $f : J \to \mathbb{R}$ is a real function defined on an interval J, that a is a particular number in J, that n is a positive

Integration with Complex Numbers. McCluskey and McMaster, Oxford University Press.
© Brian McMaster and Aisling McCluskey (2022). DOI: 10.1093/oso/9780192846075.003.0008

integer and that f is differentiable at least $n + 1$ times throughout J. Then for any x in J we have

$$f(x) = a_0 + a_1(x - a) + a_2(x - a)^2 + a_3(x - a)^3 + \cdots + a_n(x - a)^n + R_n(x),$$

where the coefficients a_j are given by the formula[1]

$$a_j = \frac{f^{(j)}(a)}{j!} \quad (0 \le j \le n)$$

and the (so-called) *nth-stage remainder* $R_n(x)$ is given by a number of different formulae, one of which is

$$R_n(x) = \frac{f^{(n+1)}(\xi)}{(n + 1)!}(x - a)^{n+1}$$

for some number ξ lying between a and x.

This is a result that has good consequences. If we are able to show that $R_n(x)$ tends to zero as $n \to \infty$, then it tells us that $f(x)$ is the sum of the power series $\sum_0^\infty \frac{f^{(n)}(a)}{n!}(x - a)^n$, which provides the means of establishing series expressions for e^x, $\sin x$, $\cos x$ and so on that we have used before (usually in the tidy special case where $a = 0$). There are, however, also some difficulties associated with 8.1.1. For one thing there is a large amount of detail to remember and to check out before applying it in any example; for another, the vagueness as to where exactly ξ is can be an irritant, and so can the fact that there is more than one potentially useful formula for $R_n(x)$. The real thorn in our collective side, though, is that *sometimes $R_n(x)$ does not tend to zero* and, in that circumstance, the theorem tells us nothing useful at all (see again paragraph 4.4.3).

What we shall now show is that, once we switch from real numbers to complex, all the 'bad' aspects of Taylor simply disappear. We only need a complex function to be differentiable *once* (on an open disc) since then all its higher derivatives will automatically exist. The formula for the coefficients does not change (although there is an interesting and valuable alternative one). The *n*th-stage remainder will *always* tend to zero so we won't really need any formula for it. Taylor effectively evolves into an if-and-only-if result: *the power series that converge within an open disc and the functions that are differentiable in that disc are the same things...* but we are getting ahead of our evidence base.

---

[1] Note the notation we are using for the jth derivative of f here. Also that $0! = 1$ and that $f^{(0)}$ just means the original function f.

8.2 Taylor series

We'll start with a tiny lemma that is going to save us time in the main proof.

8.2.1 Lemma Suppose that a is some particular point in the complex plane, and that w and z are distinct from a and from one another, and that we know which of them is further away from a: say, $|z - a| < |w - a|$. Then

$$\frac{1}{w - z} = \sum_{0}^{\infty} \frac{(z - a)^n}{(w - a)^{n+1}}$$

the series being an (absolutely convergent) geometric series.

Proof Write $w - z$ as $(w - a) - (z - a)$ and, in turn, as $(w - a)\left(1 - \frac{z - a}{w - a}\right)$. Then we have

$$\frac{1}{w - z} = (w - a)^{-1}\left(1 - \frac{z - a}{w - a}\right)^{-1} = \frac{1}{w - a}\sum_{0}^{\infty}\left(\frac{z - a}{w - a}\right)^n$$

$$= \sum_{0}^{\infty} \frac{(z - a)^n}{(w - a)^{n+1}} \quad \text{as predicted.}$$

The series is absolutely convergent because the modulus of its common ratio

$$\left|\frac{z - a}{w - a}\right| = \frac{|z - a|}{|w - a|}$$

is less than 1. (The reader who, perfectly reasonably, wonders why on earth anyone would want to replace such a simple piece of algebra as $\frac{1}{w-z}$ by a whole infinite series will see at least part of the answer in the upcoming proof.) ∎

8.2.2 Note If a function $f(z)$ can be expressed as the sum of a power series $\sum_{0}^{\infty} a_n(z - a)^n$ on an open disc $D = D(a, r) = \{z \in \mathbb{C} : |z - a| < r\}$, then we know from term-by-term differentiation that

1. it can be differentiated as often as we wish (because each differentiation just produces another absolutely convergent power series), and

2. there is no ambiguity as to what the coefficients of that power series are: because as we run the differentiation process time after time:

$$f(z) = a_0 + a_1(z-a) + a_2(z-a)^2 + a_3(z-a)^3 + a_4(z-a)^4 + a_5(z-a)^5 + \cdots \Rightarrow f(a) = a_0,$$

$$f'(z) = a_1 + 2a_2(z-a) + 3a_3(z-a)^2 + 4a_4(z-a)^3 + 5a_5(z-a)^4 + \cdots \Rightarrow f'(a) = a_1,$$

$$f''(z) = 2a_2 + 6a_3(z-a) + 12a_4(z-a)^2 + 20a_5(z-a)^3 + \cdots \Rightarrow f''(a) = 2a_2 = 2!a_2,$$

$$f^{(3)}(z) = 6a_3 + 24a_4(z-a) + 60a_5(z-a)^2 + \cdots \Rightarrow f^{(3)}(a) = 6a_3 = 3!a_3,$$

$$f^{(4)}(z) = 24a_4 + 120a_5(z-a) + \cdots \Rightarrow f^{(4)}(a) = 24a_4 = 4!a_4$$

and so on, the pattern emerging is that a_n must be the number $\dfrac{f^{(n)}(a)}{n!}$ for each non-negative integer n.

The word to focus attention on in this note is, as so often happens in mathematics, the word 'if' right at the beginning. At this point in our account, we have no guarantee that all the useful and important functions actually can be expressed as sums of power series. Fortunately, all that is about to change.

8.2.3 Taylor's theorem (complex case)

Any function that is regular on an open disc (centre $a \in \mathbb{C}$) can be expressed as a power series (in powers of $z - a$).

Proof Suppose that f is regular on $D = D(a, r)$. We take a typical point z in D, and choose a circle C (centred on a) whose radius is greater than $|z - a|$ (but less than r), that is, so that z lies inside C. From Cauchy's integral formula (and dividing by $2\pi i$ for convenience),

$$f(z) = \frac{1}{2\pi i} \int_C \frac{f(w)}{w - z} \, dw.$$

Here we replace $\dfrac{1}{w - z}$ by the power series indicated in 8.2.1

$$f(z) = \frac{1}{2\pi i} \int_C f(w) \sum_0^\infty \frac{(z-a)^n}{(w-a)^{n+1}} \, dw$$

and then *we do some magic by swopping over the sigma sign and the integral.*[2] That gives us

$$f(z) = \frac{1}{2\pi i} \sum_0^\infty \int_C f(w) \frac{(z-a)^n}{(w-a)^{n+1}} \, dw.$$

---

[2] It is not always correct to interchange summation and integral in this manner, so there is a serious gap in the argument at this point; to see how to bridge it, the reader may want to read paragraph 11.3.15 here.

Now while w traverses the circle C, neither z nor a varies with w so, from the point of view of the integral, $(z - a)^n$ is effectively a constant and can be factored out of the integration, giving

$$f(z) = \sum_0^\infty \left(\frac{1}{2\pi i} \int_C \frac{f(w)}{(w - a)^{n+1}} \, dw \right) (z - a)^n.$$

The expression in parentheses is in effect a constant (for each value of n), that is, it is independent of the choice of C: because if we alter C to another circle C' centred on a and of radius less than r, then paragraph 7.6.5 tells us that the integrals around C and around C' will be equal. Denoting it then by a_n, we see that

$$f(z) = \sum_0^\infty a_n(z - a)^n$$

which is what we wanted, since z was any point in the disc. ∎

8.2.4 Postscript

When f (regular on $D(a, r)$) is expressed as a power series $\sum_0^\infty a_n(z - a)^n$, then we have two alternative formulae for the constants a_n (the Taylor coefficients): for each $n \geq 0$,

-

$$a_n = \frac{f^{(n)}(a)}{n!} \quad \text{and}$$

-

$$a_n = \frac{1}{2\pi i} \int_C \frac{f(w)}{(w - a)^{n+1}} \, dw$$

where C is any circular path (anticlockwise, single orbit) centred on a and of radius less than r.

Although the first of these two formulae for the Taylor coefficients looks easier, the second can be very powerful in the right circumstances. Here as illustration is a result we promised back in Chapter 4 (see paragraph 4.4.6):

8.2.5 Liouville's theorem

The only functions that are both bounded and regular[3] everywhere in $\mathbb{C}$ are the constant functions.

Proof Suppose that f is regular on $\mathbb{C}$ and that we find a constant K such that $|f(z)| \leq K$ for all values of z. Taylor assures us that $f(z)$ can, for all z, be written as a power series

---

[3] A function that is regular on the whole of $\mathbb{C}$ is usually called an *entire function*.

$$f(z) = a_0 + a_1z + a_2z^2 + a_3z^3 + \cdots + a_nz^n + \cdots$$

(we are taking $a = 0$ as the centre of the expansion for convenience of notation) and that

$$a_n = \frac{1}{2\pi i} \int_C \frac{f(w)}{w^{n+1}} \, dw.$$

We can make the circular path of the integral as big as we please: say, the circle $|z| = R$ where the radius R is very large. By the 'estimation of the integral' theorem 6.6.16,

$$|a_n| \leq \left| \frac{1}{2\pi i} \right| \left(\frac{K}{R^{n+1}} \right) 2\pi R = \frac{K}{R^n}.$$

Yet $K/R^n \to 0$ as $R \to \infty$ for each $n \geq 1$, and a_n is independent of R. This forces $0 = a_1 = a_2 = a_3 = \cdots$ so the power series expansion collapses to a constant $f(z) = a_0 + 0 + 0 + 0 + 0 + \cdots$.   ∎

As a post-postscript to Taylor, we get an interesting connection between multiple differentiating and integrating, just by combining 8.2.4's two formulae for the Taylor coefficients:

8.2.6 The 'derivatives formula'   When f is regular on $D(a, r)$ and n is any non-negative integer, then

$$f^{(n)}(a) = \frac{n!}{2\pi i} \int_C \frac{f(w)}{(w - a)^{n+1}} \, dw$$

for any circle C centre a and of radius less than r. (Notice that the special case $n = 0$ is really just Cauchy's integral formula again.)

8.3 Examples

8.3.1 Example   Obtain the standard power series expansions for the complex functions $\sin z$, $\cos z$ and e^z based on their well-known derivatives.

Partial solution   This is an entirely routine exercise on the use of Taylor's theorem once one has differentiated each of these functions far enough to see the pattern of their nth derivatives evaluated at zero.

8.3.2 Example   Find a power series expansion for the function $f(z) = \dfrac{e^z}{1 - z}$ valid on the open disc $D(0, 1)$.

Solution   Since f is regular on $D(0,1)$, Taylor's theorem guarantees that such a power series must exist. In principle we could find its coefficients by repeatedly differentiating f and using the first formula in 8.2.4, but it is easier to build on things we know already:

$$f(z) = e^z(1-z)^{-1}$$

$$= \left(1 + \frac{z}{1!} + \frac{z^2}{2!} + \frac{z^3}{3!} + \cdots + \frac{z^n}{n!} + \cdots\right)(1 + z + z^2 + z^3 + \cdots + z^n + \cdots)$$

$$= 1 + \left(1 + \frac{1}{1!}\right)z + \left(1 + \frac{1}{1!} + \frac{1}{2!}\right)z^2 + \left(1 + \frac{1}{1!} + \frac{1}{2!} + \frac{1}{3!}\right)z^3 + \cdots$$

$$= \sum_0^\infty a_n z^n \text{ where (for each } n \geq 0\text{):}$$

$$a_n = 1 + \frac{1}{1!} + \frac{1}{2!} + \frac{1}{3!} + \cdots + \frac{1}{n!}.$$

(Paragraph 2.3.6 flagged up the fact that we can multiply together absolutely convergent infinite series in this fairly natural manner, known as the *Cauchy product property* of series.)

8.3.3 Example   Find, as a power series, a function f regular on the unit disc $D(0,1)$ that satisfies $f(0) = f'(0) = 1$ and, for every $n \geq 1$:

$$f^{(n+1)}(0) = \left(\frac{2n+1}{2n+3}\right)f^{(n)}(0).$$

Solution   Assuming for the moment that any such function exists, Taylor tells us that it can only be

$$f(z) = 1 + z + \frac{f^{(2)}(0)}{2!}z^2 + \frac{f^{(3)}(0)}{3!}z^3 + \frac{f^{(4)}(0)}{4!}z^4 + \cdots.$$

Now $f^{(2)}(0) = \frac{3}{5}f^{(1)}(0) = \frac{3}{5}$, $f^{(3)}(0) = \frac{5}{7}f^{(2)}(0) = \frac{3}{7}$, $f^{(4)}(0) = \frac{7}{9}f^{(3)}(0) = \frac{3}{9}$, and the emerging pattern is that $f^{(n)}(0) = \dfrac{3}{2n+1}$ for each $n \geq 2$. So the power series that defines f is

$$f(z) = 1 + z + \frac{3}{5(2!)}z^2 + \frac{3}{7(3!)}z^3 + \frac{3}{9(4!)}z^4 + \cdots + \frac{3}{(2n+1)n!}z^n + \cdots.$$

(It is an easy application of the ratio test to check that this series is absolutely convergent not only on $D(0,1)$ but on $\mathbb{C}$ itself, so the desired function does indeed exist.)

8.3.4 Example Let C denote the circle $C(0, 5)$ as a simple closed contour (transcribed anticlockwise). Evaluate

$$\int_C \frac{z^3 + 2z^2 + 3z}{(z - 2 - 3i)^3} \, dz.$$

Solution Putting $a = 2 + 3i$ and $f(z) = z^3 + 2z^2 + 3z$ we note that a lies inside the circle C (although it is not the centre). Take a small circle C_1 centred on a and contained in the interior of C. By Lemma 7.6.1 (or 7.6.5) the integrals of $\dfrac{f(z)}{(z - a)^3}$ around C and around C_1 are equal, and now using the derivatives formula 8.2.6:

$$\int_C \frac{f(z)}{(z - a)^3} \, dz = \int_{C_1} \frac{f(z)}{(z - a)^3} \, dz = \frac{2\pi i}{2!} f^{(2)}(a) = \pi i(6a + 4) = \pi(-18 + 16i).$$

8.3.5 Example Let Q denote the square whose vertices are the four points $\pm 1 \pm i$ and regarded as a simple closed anticlockwise contour. Evaluate

$$\int_Q \frac{\sin(3z) + e^{2z}}{z^{100}} \, dz.$$

Solution Setting $a = 0$ and $f(z) = \sin(3z) + e^{2z}$, notice that a lies inside Q. Take a small circle C centred on a and contained in the interior of Q. By Lemma 7.6.1 or 7.6.5, the integrals of $\dfrac{f(z)}{z^{100}}$ around Q and around C are equal. Now 8.2.6 yields

$$\int_Q \frac{f(z)}{z^{100}} \, dz = \int_C \frac{f(z)}{z^{100}} \, dz = \frac{2\pi i}{99!} f^{(99)}(a).$$

Now with a little care we can see that the 99th derivative of $\sin(3z) + e^{2z}$ is $-3^{99} \cos(3z) + 2^{99} e^{2z}$. The given integral therefore evaluates as

$$\frac{2(2^{99} - 3^{99})\pi i}{99!}.$$

8.3.6 Example If f is an entire[4] function and $|f(z)| \leq 1000|e^z|$ for every complex number z, show that f must actually be a constant multiple of the exponential function throughout $\mathbb{C}$.

Solution Since e^z is entire and never takes the value zero, the function $g(z) = \dfrac{f(z)}{e^z}$ is also entire. Also $|g(z)| \leq 1000$ so g is bounded throughout $\mathbb{C}$. By Liouville 8.2.5 it must be constant: $g(z) = \dfrac{f(z)}{e^z} = K$ for all $z \in \mathbb{C}$ and some constant K. Hence the result.

---

[4] That is, regular on the whole of $\mathbb{C}$.

8.4 Zeros

A *zero* of a regular function f means a point in $\mathbb{C}$ at which f takes the value 0. For instance, each integer is a zero for the function $\sin(\pi z)$, and the zeros of $(z^2 + 1)^2$ are i and $-i$.

8.4.1 **Note and definitions**   Let f be regular on a domain D and let a (in D) be a zero of f. Since D is an open set, we can find an open disc $D(a, r)$ centred on a and small enough to fit inside D. By Taylor, f can be expressed as a power series within $D(a, r)$ (and possibly further out in D also):

$$f(z) = a_0 + a_1(z - a) + a_2(z - a)^2 + a_3(z - a)^3 + \cdots.$$

Setting $z = a$, we find that $f(a) = a_0 + 0 + 0 + 0 + \cdots$, that is, $f(a) = a_0$. Therefore $a_0 = 0$ since a is one of f's zeros. There are then only two possibilities for the later coefficients:

1. either *all* of the a_n's are zero,
2. or there is an 'earliest' coefficient in the expansion that is not zero; it will be a_m for some integer $m \geq 1$. Then we shall have $0 = a_0 = a_1 = a_2 = \cdots = a_{m-1}$, but $a_m \neq 0$.

Look at the consequences for the function f of these two scenarios.

In case (1), $f(z)$ takes the value 0 not just at $z = a$, but everywhere throughout the disc $D(a, r)$. This is not impossible, but surely a function behaving like this will be unlikely to hold much interest for us. (We shall later show that such an f must be zero everywhere in the whole domain D, not just in the little disc.)

In case (2), more interestingly, we get

$$f(z) = 0 + 0 + 0 + \ldots + 0 + a_m(z - a)^m + a_{m+1}(z - a)^{m+1} + a_{m+2}(z - a)^{m+2} + \cdots$$

and it will pay us to tidy that up in an informative way:

$$f(z) = (z - a)^m (a_m + a_{m+1}(z - a) + a_{m+2}(z - a)^2 + \cdots).$$

The longer expression in parentheses is the sum of another power series *and is therefore a regular function*[5]—call it $g(z)$, for instance—and notice that $g(a) = a_m \neq 0$.

---

[5] We need to keep in mind the essence of what Taylor's theorem tells us: that regular functions and convergent power series are virtually the same thing!

In this case we say that f *has a zero of order m at a*. What we have just now demonstrated about it is a kind of factorization lemma:

8.4.2 Factorization at a zero of finite order   If f is regular on a disc $D(a, r)$ and has a zero of order m at a then, on that disc,

$$f(z) = (z - a)^m g(z)$$

for some regular function g that does not have a zero at a.

The converse is also true: for if g is regular but non-zero at a, then its Taylor expansion begins with $a_0 + a_1(z - a) + \cdots$ where $a_0 \neq 0$; therefore

$$(z - a)^m g(z) = a_0(z - a)^m + a_1(z - a)^{m+1} + \cdots$$

has a zero of order m at a.

8.4.3 Note   Zeros of orders 1, 2, 3... are called *simple, double, triple*... and you may occasionally meet terms like *quadruple* or *quintuple* zeros (but not often).

8.4.4 Examples

1. Any function f that is regular on a disc centre a and has $f(a) = 0$ but $f'(a)$ non-zero must have a simple zero at a: because its order-1 Taylor coefficient at a (namely $a_1 = \dfrac{f^{(1)}(a)}{1!}$) is already non-zero. So, for instance, $\sin z$ and $z e^z$ and $z + 2z^2 + 5z^4$ have simple zeros at 0, $\cos(\pi z)$ has a simple zero at $\frac{1}{2}$, and $1 - \tan z$ has a simple zero at $\frac{\pi}{4}$.

2. The function $z \sin z$ has the Taylor expansion

$$z^2 - \frac{z^4}{3!} + \frac{z^6}{5!} - \frac{z^7}{6!} + \cdots$$

and therefore has a double zero at 0. So does

$$1 - \cos z = \frac{z^2}{2!} - \frac{z^4}{4!} + \frac{z^6}{6!} - \cdots .$$

3. The function $(z^2 + 100)^3$ has a triple zero at $10i$ because of its factorization

$$(z^2 + 100)^3 = (z - 10i)^3 g(z)$$

where $g(z) = (z + 10i)^3$ is regular and non-zero at $10i$ (note the converse implication in 8.4.2). Likewise, it has another triple zero at $-10i$.

8.4.5 **The isolated zeros theorem**   If a regular function f has a zero of finite order at a point a, then there is a disc centred on a within which a is the *only* zero of f.

Proof   Let us start with f being regular on a domain D and possessing a zero of order m at $a \in D$. Since D is open, we can choose a first disc $D(a, r)$ centred on a and contained in D. By 8.4.2, f factorizes on $D(a, r)$: $f(z) = (z - a)^m g(z)$ where g is also regular and $g(a) \neq 0$. Since it is regular, g is certainly continuous, so we can find a positive δ (smaller than r) such that $|z - a| < \delta$ forces $|g(z) - g(a)| < |g(a)|$ which in turn guarantees that $g(z) \neq 0$. So g is never zero in the smaller disc $D(a, \delta)$ and, of course, $(z - a)^m$ is zero only at a. Therefore a is the only place in $D(a, \delta)$ at which $f(z) = (z - a)^m g(z)$ can take the value 0.   ∎

8.4.6 **Remark**   This is another point at which real analysis and complex analysis behave in strikingly different ways. Although the zeros of a (non-constant) regular (that is, differentiable) *complex* function have to be spaced out from one another as we have just seen, no such self-restraint is observed in differentiable *real* functions. For instance, the real function f described by

$$f(x) = \begin{cases} x^2 \sin\left(\dfrac{\pi}{x}\right) & \text{if } x \neq 0, \\ 0 & \text{if } x = 0 \end{cases}$$

is well known to be differentiable throughout $\mathbb{R}$, and it has zeros at 0 and at $\frac{1}{n}$ for every non-zero integer n. It is true that each of the $\frac{1}{n}$ is 'isolated' by having some clear space on each side of it containing no other zero, but any attempt to find an open interval centred on 0 that will incorporate no other zero of f is evidently doomed to fail, since $\frac{1}{n} \to 0$ as $n \to \infty$.

Although it is a digression from our primary purpose, we are now so close to two major theorems that it would be a shame not to visit them:

8.4.7 **The identity theorem, version 1**   Suppose that f is regular on a domain D, and that we can find (in D) a zero a of f that is *not* an isolated zero. Then f is zero everywhere[6] throughout D.

Proof   Suppose, if possible, that there is some point b in D where $f(b)$ is not zero (and now we seek a **contradiction**). From 8.4.1 we know that there must be a disc $D(a, r)$ on which f is constantly zero. Also, since D is connected, there has to be a path $\gamma : [0, 1] \to D$ within D from a to b. Informally, the strategy is to travel along γ as far as we can go without finding a place where f is not zero, and then to inspect carefully the position at which we are forced to stop.

---

[6] Often worded as f is *identically zero* on D.

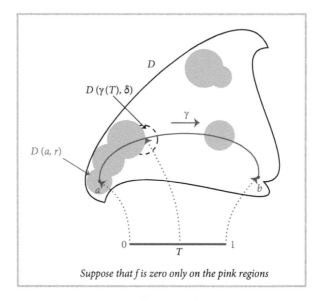

Suppose that f is zero only on the pink regions

Formally, let T be the supremum of the set $\{t \in [0,1] : f$ is identically zero on $\gamma([0,t])\}$. Then T is strictly greater than 0 (since γ begins its journey through $D(a,r)$). Therefore there are values of t arbitrarily close to but less than T at which $f(\gamma(t))$ is zero. Using continuity, this tells us that $f(\gamma(T))$ must also be zero, but that $\gamma(T)$ cannot be an *isolated* zero of f. Furthermore, $\gamma(T)$ cannot equal b since $f(b) \neq 0$. Appealing to 8.4.1 a second time, there must be a disc $D(\gamma(T), \delta)$, centred on $\gamma(T)$ this time, throughout which f is again constantly zero. Yet this implies that there are 'times' greater than T up to which f has always been zero along γ—which successfully **contradicts** the choice of T. ∎

8.4.8 The identity theorem, version 2   Suppose that two functions f and g are regular on a domain D, and that we can find a sequence $(z_n)_{n \in \mathbb{N}}$ of distinct points of D that converges to a limit in D, and for which $f(z_n) = g(z_n)$ for every n. Then f and g are identically the same function throughout D.

Proof   Just apply version 1 to the function $f - g$. ∎

8.4.9 Note   The same example that we used in 8.4.6 shows that, once again, real analysis behaves differently from complex: for the function f that we described there coincides with the constantly zero function at every stage along the convergent sequence $(1/n)$, and yet f is certainly not identically the same as the constantly zero function.

8.4.10 The fundamental theorem of algebra   Every polynomial[7] equation of degree greater than zero has a root in $\mathbb{C}$.

---

[7] With real or complex coefficients.

That is, if $p(z) = a_0 + a_1 z + a_2 z^2 + \cdots + a_{n-1} z^{n-1} + a_n z^n$ (where $a_n \neq 0$) is any polynomial of degree $n \geq 1$, then there is a complex number z_1 such that $p(z_1) = 0$.

Proof   For very big values of $|z|$, the term $a_n z^n$ in $p(z)$ is going to be more important than all the other terms put together. The first step in the proof is to take that informal insight and clarify it.

- There is a circle $|z| = L$ *outside of which* $0.9|a_n z^n| \leq |p(z)| \leq 1.1|a_n z^n|$ $\cdots$

    $\ldots$ for if we express $p(z)$ as $z^n \left(a_n + \dfrac{a_{n-1}}{z} + \dfrac{a_{n-2}}{z^2} + \dfrac{a_{n-3}}{z^3} + \cdots + \dfrac{a_0}{z^n} \right)$ and
    notice that the expression in parentheses (let's call it $g(z)$) tends to a_n as $|z| \to \infty$,
    then we can force $|g(z)|$ to lie between $0.9|a_n|$ and $1.1|a_n|$ just by taking $|z|$
    sufficiently big… say, $|z| \geq L$. Hence the result.
    In particular, everywhere outside the circle $|z| = L$ we get

    $$|p(z)| \geq 0.9|a_n z^n| \geq 0.9|a_n|L^n.$$

- Now the rational function $\dfrac{1}{p(z)}$ is differentiable everywhere except at any places
    where the bottom line takes the value zero. *If $p(z)$ never takes the value zero*
    then $\dfrac{1}{p(z)}$ is regular everywhere in $\mathbb{C}$ and also, being a continuous function, it is
    bounded on the bounded closed disc $\overline{D}(0, L)$. Yet we also know from the previ-
    ous display that $\dfrac{1}{p(z)}$ cannot in modulus exceed $\dfrac{1}{0.9|a_n|L^n}$ outside $D(0, L)$, and
    therefore $\dfrac{1}{p(z)}$ is bounded everywhere in $\mathbb{C}$. By Liouville (8.2.5) it has to be a
    constant, and so must $p(z)$ itself. *This is a contradiction* because p has degree 1
    or more.
    Therefore $p(z)$ must take the value zero somewhere in the complex plane. ∎

8.4.11 **Postscript**   Continuing in the same notation, now that we know $p(z)$
has a zero of some (finite) order at z_1, 8.4.2 tells us that it must factorise as $p(z) = (z - z_1)^m g(z)$, where $g(z)$ is regular and is therefore the sum of a power series. By
inspection, that power series can only be another polynomial of degree smaller
than n. Using 8.4.10 and the same argument on $g(z)$ in its turn, it must take the
form $(z - z_2)^p h(z)$ where $h(z)$ can only be a polynomial of still smaller degree.
Continuing this process until we eventually reach a polynomial of zero degree,
that is, a non-zero constant, we conclude that every polynomial $p(z)$ 'factorises
completely over the complex numbers', that is, we shall get

$$p(z) = K(z - z_1)^m (z - z_2)^p (z - z_3)^q \cdots$$

where $K \neq 0$ is a constant and the orders of the zeros add up to the degree of p.
This is the conclusion that is often expressed as 'every (real or complex) polynomial of degree n has exactly n (complex) roots *counted according to multiplicity*'.

8.5 Exercises

Throughout this set of exercises, a symbol of the form $C(z_0, r)$ represents the circle of centre z_0 and radius r, as a simple closed contour of integration transcribed once anticlockwise, starting and ending at the point $z_0 + r$.

1. Starting from the standard (Taylor) series expansions of $\sin z$, $\cos z$ and e^z at $z = 0$, write down the values of the following integrals (in which C denotes a small anticlockwise circular contour centred on the origin):

   (a)
   $$\int_C \frac{\sin z}{z^4}\, dz \qquad \int_C \frac{\sin z}{z^7}\, dz \qquad \int_C \frac{\sin z}{z^{10}}\, dz$$

   (b)
   $$\int_C \frac{\cos z}{z^3}\, dz \qquad \int_C \frac{\cos z}{z^7}\, dz \qquad \int_C \frac{\cos z}{z^8}\, dz$$

   (c)
   $$\int_C \frac{e^z}{z^2}\, dz \qquad \int_C \frac{e^z}{z^6}\, dz \qquad \int_C \frac{e^z}{z^9}\, dz$$

2. Use the derivatives formula to evaluate the following integrals (where, in each case, the contour of integration is a small anticlockwise simple closed circular path centred on the zero of the bottom line of the integrand):

   (a) $\int_C \dfrac{e^{2z}}{(z-3)^2}\, dz$

   (b) $\int_C \dfrac{(z^3+1)^3}{(z+3)^3}\, dz$

   (c) $\int_C \dfrac{ze^z}{(z+i)^4}\, dz$

3. Find a power series representing a function f regular on the unit disc $D(0,1)$ that satisfies $f(0) = 3, f'(0) = 2$ and, for every $n \geq 1$:

   $$f^{(n+1)}(0) = \left(\frac{3n-1}{3n+2} \right) f^{(n)}(0).$$

4. For the function f identified in Exercise 3, evaluate

   $$\int_C \frac{f(z)}{z^{35}}\, dz, \qquad \int_C \frac{f(z)}{(1-z)z^4}\, dz, \qquad \int_C \frac{(f(z))^2}{z^3}\, dz$$

   where $C = C(0, 0.1)$.

5. Derive the Taylor expansion of the function $f(z) = ze^{2z}$ valid on a disc centred at $z = 1$, that is, as a series of powers of $(z-1)$.

Using this series, evaluate the integrals

$$\int_{C(1,1)} \frac{f(z)}{(z-1)^{13}}\,dz \quad \text{and} \quad \int_{C(1,1)} \frac{f(z)\cos(z-1)}{(z-1)^4}\,dz.$$

6. Recall from Exercise 22 of Chapter 4 that the function $\operatorname{Log} z$ (the principal value of the complex logarithm) is defined on the cut plane $\mathbb{C}\setminus(-\infty,0]$ and is regular there, with derivative $\dfrac{1}{z}$, and that $e^{\operatorname{Log} z}=z$ for all relevant z.

   Obtain the Taylor series for this function on the disc $D(1,1)$, that is, expressing $\operatorname{Log}(1+w)$ as a power series in w for $|w|<1$. Use it to evaluate

$$\int_K \frac{\operatorname{Log} z}{(z-1)^3}\,dz, \quad \int_K \frac{\operatorname{Log} z}{(z-1)^6}\,dz \quad \text{and} \quad \int_K \frac{e^z\,\operatorname{Log} z}{(z-1)^4}\,dz$$

   where K is $C(1,0.5)$.

7. Continuing from Exercise 6, obtain an approximation to $\operatorname{Log}\left(1+\dfrac{i}{20}\right)$ aiming for correctness to six decimal places. Check informally for accuracy by substituting your answer into the exponential function e^z.

8. The principal value of the square root function, usually just written as $z^{\frac12}$, is defined on the cut plane $\mathbb{C}\setminus(-\infty,0]$ by

$$z^{\frac12} = e^{(\operatorname{Log} z)/2}.$$

   Check that it is regular on the cut plane, and obtain the first six terms of a Taylor expansion of this function valid on the disc $D(1,1)$, that is, expressing

$$(1+w)^{\frac12} = a_0 + a_1 w + a_2 w^2 + a_3 w^3 + a_4 w^4 + a_5 w^5 \cdots.$$

9. Use the series developed in Exercise 8 to find an approximate value of

$$\left(1+\frac{1+i}{10\sqrt2}\right)^{\frac12}$$

   aiming for correctness to six decimal places. Informally check for accuracy by substituting your answer into the function z^2.

10. Let $f(z)$ be an entire function and suppose that the imaginary part of $f(z)$ is bounded above. Prove that $f(z)$ is actually constant.

    How would you modify the demonstration to show that an entire function whose imaginary part is bounded below, or whose real part is bounded above, or whose real part is bounded below, must be constant?

11. Suppose that an entire function $f(z)$ and a polynomial $p(z)$ satisfy the condition

$$|f(z)| \le |p(z)| \quad \text{for every } z \in \mathbb{C}.$$

Use Taylor's theorem to show that $f(z)$ must be a polynomial. (*Suggestion:* reconsider the proof of Liouville's theorem.)

12. Identify the zeros of the following functions and the order of each zero:
    (a) $(z^3 + 1)(z - 1)^5$,
    (b) $\sin(\pi z)$,
    (c) $e^{\cos z}\left(1 + \dfrac{3}{z^4} + \dfrac{3}{z^8} + \dfrac{1}{z^{12}}\right)$,
    (d) $\sin(\pi e^z)$.

13. Evaluate the following integrals:
    (a) $\displaystyle\int_C \frac{3z^3 - 5z^2 + 7z - 9}{z^2 - 6z + 25}\, dz$ where $C = C(3 + 4i, 1)$,
    (b) $\displaystyle\int_Q \frac{2\cos(\pi z) + e^{-z}}{z^{55}}\, dz$ where Q is the simple, closed, anticlockwise contour formed by the rectangle having vertices at $1 - i, 1 + i, -2 + i$ and $-2 - i$,
    (c) $\displaystyle\int_C \frac{\sin\left(e^{\sin z} - 1\right)}{z^4}\, dz$ where $C = C(0, 0.01)$.

14. (a) What can be guaranteed about a function f that is regular on the disc $D(0, 1)$ and satisfies, for each positive integer n, the condition

$$f\left(\frac{1}{n+1}\right) = \frac{1}{(n+1)^2}?$$

(b) What can be guaranteed about a function f that is regular on the disc $D(0, 1)$ and satisfies, for each positive integer n, the condition

$$f\left(\frac{1}{n+1}\right) = \frac{(-1)^n}{(n+1)^2}?$$

(c) Can we formulate any equally definite conclusion about a function f that is regular on the disc $D(0, 1)$ and satisfies, for each positive integer n, the condition

$$f\left(\frac{n}{n+1}\right) = \frac{n^2}{(n+1)^2}?$$

15. (a) Let $p(z)$ be a polynomial of degree n that possesses fewer than n distinct roots in $\mathbb{C}$. Show that there is $w \in \mathbb{C}$ such that both $p(w) = 0$ and $p'(w) = 0$.

(b) Conversely, if $p(z)$ is a polynomial of degree n and some number is simultaneously a root of $p(z)$ and a root of its (first) derivative $p'(z)$, show that $p(z)$ has fewer than n distinct roots in $\mathbb{C}$.

(c) Is it possible for a polynomial $p(z)$ of degree n to have n distinct roots and yet for it and its *second* derivative $p''(z)$ to have a root in common? (*Suggestion:* it is enough to consider the case $n = 3$.)

(d) (For $n \in \mathbb{N}$) show that the equation $z^n + z^{n-1} + z^{n-2} + \cdots + z^2 + z + 1 = 0$ has n distinct roots.

9 Residues

9.1 Laurent's theorem

Taylor's theorem showed us that a function f that is regular on a disc D centred at a can be expressed as a power series in $z - a$. What, then, could we say if regularity failed at a but not elsewhere in the disc D? It is not an unreasonable question, because many otherwise well-behaved functions fail differentiability (or even fail to be defined) at a handful of awkward points. For instance, it could be useful to understand how $\dfrac{\sin z}{z}$ behaves close to zero, or how $\dfrac{1}{z^2 + 1}$ behaves close to (but evidently not exactly at) i and at $-i$.

The key tool in such investigations is a result called Laurent's theorem. Roughly speaking, it says that these functions can also be expressed as 'power series' in $z - a$ *but* we have to allow negative powers of $z - a$ in addition to the non-negative powers that Taylor handles. Both the result itself and its proof run very much parallel to those of Taylor except, of course, that we cannot this time expect the nth derivative of f at a to help us work out the nth coefficient (since f could very well not even be defined at a, let alone n times differentiable there).

9.1.1 Laurent's theorem   Suppose that the function f is regular on the disc $D = D(a, L)$ except at a, its centre.[1] Then at every point z of the punctured disc $D \setminus \{a\}$ we can express $f(z)$ as the sum of a 'double power series'

$$f(z) = \sum_{-\infty}^{\infty} a_n (z - a)^n$$

where the Laurent coefficients a_n for each $n \in \mathbb{Z}$ are given (as in Taylor's theorem) by

$$a_n = \frac{1}{2\pi i} \int_K \frac{f(w)}{(w - a)^{n+1}} \, dw$$

and K is any circle centred on a and of radius less than L.

---

[1] The whole result is also valid if f happens to be regular at a too—but in that case, you would have no need to use Laurent, you'd get whatever you wanted from Taylor's theorem instead.

Integration with Complex Numbers. McCluskey and McMaster, Oxford University Press.
© Brian McMaster and Aisling McCluskey (2022). DOI: 10.1093/oso/9780192846075.003.0009

Proof   Consider a typical point z in the punctured disc, and choose two circles C^+, C^- centred on a and of radii less than L such that z lies inside C^+ but outside C^-.

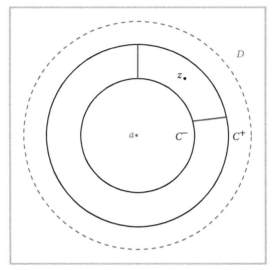

We join the inner and outer circles by two pieces of straight line (refer again to the diagram) carefully avoiding z, and notice that the function $\dfrac{f(w)}{w-z}$ is regular throughout D except at the two points a and z. Next we focus attention on the annular region bounded by C^+ and C^-, separating out the part chopped off by our *blue* construction lines that contains z (call its boundary Γ_2) from the part that does not contain z (call its boundary Γ_1). Again, please use the diagrams to keep track of what is intended here. Both Γ_2 and Γ_1 are to be transcribed anticlockwise (and just once).

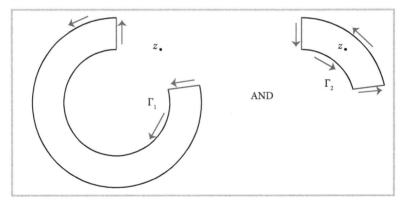

So $\dfrac{f(w)}{w-z}$ is regular on and within Γ_1 but evidently not at the point z inside Γ_2. Using Cauchy's theorem on the first and Cauchy's integral formula on the second, we learn that

$$\int_{\Gamma_1} \frac{f(w)}{w-z}\, dz = 0 \quad \text{and} \quad \int_{\Gamma_2} \frac{f(w)}{w-z}\, dz = 2\pi i\, f(z).$$

When we add these, notice that the back-and-forth integrals along the (blue) joining straight lines cancel out, and that the inner circle has been transcribed *clockwise*. Therefore we find ourselves left with

$$2\pi i\, f(z) = \int_{C^+} \frac{f(w)}{w-z}\, dw - \int_{C^-} \frac{f(w)}{w-z}\, dw$$

$= I_1 - I_2$, let us say. The first of these two integrals, I_1, we process exactly as we did in the proof of Taylor's theorem, to get

$$I_1 = \sum_{n=0}^{\infty} \left(\int_{C^+} \frac{f(w)}{(w-a)^{n+1}}\, dw \right) (z-a)^n.$$

The second is reshaped in just the same fashion,[2] but notice that this time the z and the w have to be swopped over: because as the controlling variable w moves around the inner circle C^-, it will be w that is closer to a than z is, so the z-to-w swop is necessary in order to use the same Lemma 8.2.1 as facilitated the proof of Taylor. This time, we get

$$-I_2 = \sum_{m=-1}^{-\infty} \left(\int_{C^-} \frac{f(w)}{(w-a)^{m+1}}\, dw \right) (z-a)^m.$$

Putting both halves together (and tidying the names of the summation variables) yields

$$f(z) = \sum_{-\infty}^{\infty} a_n (z-a)^n$$

with the coefficients as specified in the statement. Lastly, the apparent ambiguity as to whether our integrals are around C^+ or C^- or some other such circle is actually irrelevant, as paragraph 7.6.5 points out. ∎

9.1.2 Notes

1. We should take time to clarify beyond doubt the meaning of double power series: to say that

$$f(z) = \sum_{-\infty}^{\infty} a_n (z-a)^n$$

---

[2] The full details are set out in paragraph 11.3.16.

actually means that both $\sum_0^\infty a_n(z-a)^n$ and $\sum_1^\infty a_{-n}(z-a)^{-n}$ converge, and that the total of their two sums-to-infinity is $f(z)$ (for each z in the punctured disc).

2. Once again (just as in Taylor) there is a uniqueness about this expansion: if a function regular on a punctured disc is expressed in two (!) ways as the sum of a double power series

$$f(z) = \sum_{-\infty}^{\infty} a_n(z-a)^n = \sum_{-\infty}^{\infty} b_n(z-a)^n$$

then necessarily $a_n = b_n$ for every integer n. However, it is less easy to prove than the corresponding Taylor uniqueness because we cannot simply use the 'nth derivative at $z = a$' device this time. Instead we need to do *magic* by swopping the order of summation and integral again. (The reader is encouraged to refer to the details in paragraph 11.3.14 when time permits.)

3. Why this can become very useful is that there are often easy and more direct ways to find a double power series description of a function regular on a punctured disc. For instance, to do this for $f(z) = \dfrac{1 - \cos z}{z^5}$, we could use Taylor to access the power series for $\cos z$, change its sign, add 1 and divide by z^5. That gives

$$\frac{1 - \cos z}{z^5} = \frac{z^{-3}}{2!} - \frac{z^{-1}}{4!} + \frac{z}{6!} - \frac{z^3}{8!} + \cdots .$$

By item 2 above, this has to be the Laurent expansion of f about $z = 0$.

4. In a Laurent expansion

$$f(z) = \sum_{-\infty}^{\infty} a_n(z-a)^n$$

it is often important to think of the 'two series' separately (as we did during the proof), as

$$f(z) = \sum_{-\infty}^{-1} a_n(z-a)^n + \sum_{0}^{\infty} a_n(z-a)^n.$$

The second of these is called the *regular part* of the expansion (after all, if f were regular at a, this is the only part of the expansion that would be there at all) or, informally, the *Taylor part*. The first, that is, the series comprising the negative powers of $z - a$ (all the fragments that threaten to 'go infinite' as we get close to $z = a$) is termed the *principal part* (or, very informally, the *dangerous part*!). How the function behaves close to a is overwhelmingly decided by the principal part of its Laurent expansion there. In the example of item 3, the principal part was $\dfrac{1}{2z^3} - \dfrac{1}{24z}$.

5. The coefficient of $\dfrac{1}{z-a}$ in the Laurent expansion is called the *residue* of the function at a. There is a lot to say about it. (In the last example, the residue was $-\frac{1}{24}$.)

9.1.3 **Examples**   Determine the Laurent series of the following functions at the points indicated. Also identify the principal part and the residue.

1. e^z at zero.
2. $e^{1/z}$ at zero.
3. $\dfrac{z - \sin z}{z^3}$ at zero.
4. $\dfrac{z - \sin z}{z^4}$ at zero.
5. $\dfrac{z - \sin z}{z^5}$ at zero.
6. $\dfrac{z - \sin z}{z^6}$ at zero.
7. $\dfrac{1}{z^2 + 1}$ at i.
8. $\csc^2 z$ at zero.

Solutions

1. e^z at zero is something of a non-question since it is regular everywhere, so the Laurent series is merely the familiar Taylor series $1 + z + z^2/2! + z^3/3! + \cdots$.

2. The function $e^{1/z}$ is regular on $\mathbb{R} \setminus \{0\}$, so it has a Laurent expansion around 0. Replacing z by $1/z$ (while $z \neq 0$) in solution 1 above, we obtain

$$e^{1/z} = 1 + \frac{1}{z} + \frac{1}{2!z^2} + \frac{1}{3!z^3} + \frac{1}{4!z^4} + \frac{1}{5!z^5} + \cdots$$

which therefore must be this Laurent expansion. The whole thing apart from the constant term 1 involves negative powers, so it is the principal part; the residue is 1.

3. Since Taylor gives us $\sin z = z - \dfrac{z^3}{3!} + \dfrac{z^5}{5!} - \dfrac{z^7}{7!} + \dfrac{z^9}{9!} - \cdots$, we find that (provided $z \neq 0$)

$$\frac{z - \sin z}{z^3} = \frac{1}{3!} - \frac{z^2}{5!} + \frac{z^4}{7!} - \frac{z^6}{9!} + \cdots.$$

The principal part is 0: indeed, it is common practice to say that there is no principal part in this case. The residue is 0.

4. By the same argument we find that

$$\frac{z - \sin z}{z^4} = \frac{1}{3!z} - \frac{z}{5!} + \frac{z^3}{7!} - \frac{z^5}{9!} + \cdots .$$

The principal part is $\dfrac{1}{3!z}$ and the residue is $1/6$.

5. Likewise,

$$\frac{z - \sin z}{z^5} = \frac{1}{3!z^2} - \frac{1}{5!} + \frac{z^2}{7!} - \frac{z^4}{9!} + \cdots .$$

The principal part is $\dfrac{1}{3!z^2}$ and the residue is 0. It may be worth reinforcing that the residue is zero even though a principal part is present here.

6. Likewise,

$$\frac{z - \sin z}{z^6} = \frac{1}{3!z^3} - \frac{1}{5!z} + \frac{z}{7!} - \frac{z^3}{9!} + \cdots .$$

The principal part is $\dfrac{1}{3!z^3} - \dfrac{1}{5!z}$ and the residue is $-\frac{1}{120}$.

7. To explore $f(z) = \dfrac{1}{z^2 + 1}$ close to $z = i$, it helps us to see through the algebra if we substitute $w = z - i$ so that z becomes $w + i$ and $f(z)$ becomes

$$((w + i)^2 + 1)^{-1} = (w^2 + 2iw)^{-1} = \frac{1}{2iw}\left(1 + \frac{w}{2i}\right)^{-1}$$

$$= \frac{1}{2iw}\left(1 - \frac{w}{2i} + \left(\frac{w}{2i}\right)^2 - \left(\frac{w}{2i}\right)^3 + \cdots\right)$$

$$= \frac{1}{2iw} + \frac{1}{4} - \frac{w}{8i} - \frac{w^2}{16} + \cdots$$

$$= \frac{-i}{2}\frac{1}{z - i} + \frac{1}{4} + \frac{i}{8}(z - i) - \frac{1}{16}(z - i)^2 + \cdots$$

(in which we have assumed $\left|\dfrac{w}{2i}\right| < 1$ to make sure that the series converges). From this we read off that the principal part is $\dfrac{-i}{2}\dfrac{1}{z - i}$ and that the residue is $\dfrac{-i}{2}$.

8. Remembering the double-angle formulae of trigonometry will save us pain here. Since $\sin^2 z = \frac{1}{2}(1 - \cos(2z))$, the required function is

$$\frac{2}{1 - \cos(2z)} = \frac{2}{1 - \left(1 - \frac{4z^2}{2!} + \frac{16z^4}{4!} - \cdots\right)} = \frac{2}{2z^2 - \frac{2z^4}{3} + \cdots}$$

$$= \frac{1}{z^2 - \frac{z^4}{3} + \cdots} = \frac{1}{z^2}\frac{1}{1 - \frac{z^2}{3} + \cdots} = \frac{1}{z^2}\left(1 - (z^2/3 + \text{terms in } z^4 \text{ and higher})\right)^{-1}$$

$$= \frac{1}{z^2}\left(1 + (z^2/3 + \text{terms in } z^4 \text{ and higher}) + \text{more terms in } z^4 \text{ and higher}\right)$$

and here we are assuming that z is close enough to zero to force the expression $(z^2/3 + \text{terms in } z^4 \text{ and higher})$ to have modulus less than one, because we need the series to converge. From this we can already see that the principal part is $\dfrac{1}{z^2}$ and that the residue is 0. (Incidentally, paragraph 9.3.2 will provide a less clumsy way of handling residue calculations such as this one.)

We have already begun to see that 'how big or complicated' the principal part of the Laurent series is can have importance on how to work with the series. The examples we have encountered already show that it can involve an infinite number of terms, or just a finite number, or even none at all. There is standard terminology to handle and distinguish between these three cases, and it is high time we introduced it.

9.1.4 Definition When f is regular on a punctured neighbourhood $D \setminus \{a\} = \{z \in \mathbb{C} : 0 < |z - a| < r\}$ of a point a but is not defined at a, we call a a *singularity* of f. More precisely, we say that a is

1. a *removable singularity* of f if there are no terms in the principal part of its Laurent expansion there,
2. a *pole* of f if there are finitely many terms (but not none!) in the principal part of its Laurent expansion there, and
3. an *isolated essential singularity* of f if there are infinitely many (non-zero) terms in the principal part of its Laurent expansion there.
4. Furthermore, in case 2 we say that a is a pole of *order n* of f if the largest power of $\dfrac{1}{z - a}$ present in the expansion is the nth power.
5. Poles of order 1, 2 and 3 are often referred to as *simple poles, double poles* and *triple poles*. (Terms such as *quadruple* and *quintuple* will also occasionally be encountered.)

9.1.5 Note Revisiting the examples of 9.1.3 in this language, we see that

1. e^z has no singularity at zero,

2. $e^{1/z}$ has an isolated essential singularity at zero,

3. $\dfrac{z - \sin z}{z^3}$ has a removable singularity at zero,

4. $\dfrac{z - \sin z}{z^4}$ has a simple pole at zero,

5. $\dfrac{z - \sin z}{z^5}$ has a double pole at zero,

6. $\dfrac{z - \sin z}{z^6}$ has a triple pole at zero,

7. $\dfrac{1}{z^2 + 1}$ has a simple pole at i, and

8. $\csc^2 z$ has a double pole at zero.

The 'coincidence' that, on the one hand, $\dfrac{\sin z}{z}$ has (only) a *removable* singularity at zero—that is, its Laurent series consists entirely of its 'Taylor' part, as if it were actually regular at zero as well as all around zero—and, on the other hand, that $\dfrac{\sin z}{z}$ famously has a limit as z approaches its apparent singularity, is very far from being a coincidence. As we now readily confirm, these two circumstances are concomitant in general; in fact, the next theorem provides us with *two* quick and easy ways to recognise removability of a singularity:

9.1.6 Theorem Let f be regular on (at least) a punctured disc $D \setminus \{a\} = \{z \in \mathbb{C} : 0 < |z - a| < r\}$ but not defined at a. Then the following conditions are equivalent:

1. f possesses a removable singularity at a,

2. the limit $\lim_{z \to a} f(z)$ exists,

3. f is bounded on some punctured disc centred on a.

Proof

- [1 implies 2] If the singularity is removable, then we remove it: that is, we take the Taylor part of the Laurent series (all of it!) to define a 'new' function F that is regular on the whole disc and coincides with f on the punctured one. Its value at a is the limit of F and also therefore of f as we approach a.

- [2 implies 3] If the limit (call it L) does exist then we can, for instance, make $|f(z) - L| < 1$ by forcing z close enough to a: say, $0 < |z - a| < \delta$. This implies that f is bounded on the punctured disc $D(a, \delta) \setminus \{a\}$.

- [3 implies 1] Suppose that, for some choice of positive $\delta < r$, we have $|f(z)| < M$ (a constant) throughout $D(a, \delta) \setminus \{a\}$. Each of the Laurent coefficients a_{-m} in the principal part of f is given by the formula

$$a_{-m} = \frac{1}{2\pi i} \int_K f(w)(w-a)^{m-1} \, dw$$

where m is a positive integer, and we can choose the radius ε of the circle K (centred on a) to be as small as we please, and in particular smaller than δ. It follows that

$$|a_{-m}| \leq \left(\frac{1}{2\pi}\right) M\varepsilon^{m-1}(2\pi\varepsilon)$$

via the estimation of the integral theorem 6.6.16; that is, $|a_{-m}| \leq M\varepsilon^m$. Letting the radius ε of the circle K tend to zero, we realise that the coefficient must have been zero. Hence the principal part has no terms other than zeros, and the singularity is removable.

■

So how do we recognise the other kinds of singularity?

A substantial part of the answer is the pleasantly symmetrical connection between poles and zeros: for it turns out that $f(z)$ has a zero of a particular order at $z = a$ if and only if $\frac{1}{f(z)}$ has a pole of that same order there. Since zeros are fairly easy to identify and classify, this will be useful to us.

9.1.7 Lemma

1. If f has a zero of order m at a, then $1/f$ has a pole of order m at a.
2. If f has a pole of order m at a, then $1/f$ has[3] a zero of order m at a.
3. As a useful extension to point 1 above: if f has a zero of order m at a, and another (regular) function g has $g(a) \neq 0$, then g/f has a pole of order m at a.

Proof

1. Suppose f has a zero of order m at a. By 8.4.2 we can factorise it as $f(z) = (z-a)^m g(z)$ on a small disc centre a, where g is regular and $g(a) \neq 0$. Also, continuity of g tells us that g will be non-zero on a (possibly smaller still) disc centre a.

   On the smaller disc (except at its centre), $\frac{1}{f(z)} = (z-a)^{-m} \frac{1}{g(z)}$ where $1/g(z)$, being the reciprocal of a non-zero regular function, is itself regular (on the whole disc including its centre) and can therefore, via Taylor, be written as a power series: $1/g(z) = a_0 + a_1(z-a) + a_2(z-a)^2 + \cdots$ where $a_0 = \frac{1}{g(a)} \neq 0$.

---

[3] Provided we remove its removable singularity.

Then

$$\frac{1}{f(z)} = (z-a)^{-m}(a_0 + a_1(z-a) + a_2(z-a)^2 + \cdots) = \frac{a_0}{(z-a)^m} + \frac{a_1}{(z-a)^{m-1}} + \cdots,$$

recognisable as a Laurent expansion at a pole of order m.

2. Suppose f has a pole of order m at a. Then on a small disc (minus its centre a) we have

$$f(z) = \frac{c_0}{(z-a)^m} + \frac{c_1}{(z-a)^{m-1}} + \frac{c_2}{(z-a)^{m-2}} + \cdots$$

where $c_0 \neq 0$. Multiplying through by $(z-a)^m$ we find that $g(z) = (z-a)^m f(z)$ is expressed as a power series on the disc *except for its centre*, and possesses a limit (namely c_0) as $z \to a$. By 9.1.6 this is a removable singularity, so we remove it: we define $g(a)$ to mean this limit c_0, and now g is regular on the whole disc.

(Shrinking the disc if necessary to avoid any zeros of g) we now have $1/g(z)$ regular, and therefore (Taylor again) expressible as a power series:

$$\frac{1}{g(z)} = d_0 + d_1(z-a) + d_2(z-a)^2 + \cdots$$

and therefore

$$\frac{1}{f(z)} = (z-a)^m \frac{1}{g(z)} = d_0(z-a)^m + d_1(z-a)^{m+1} + d_2(z-a)^{m+2} + \cdots$$

has, indeed,[4] a zero of order m at a.

3. Supposing that f and g are as described, pick up from the proof of point 1 the Laurent expansion

$$\frac{1}{f(z)} = \frac{a_0}{(z-a)^m} + \frac{a_1}{(z-a)^{m-1}} + \cdots.$$

Also, g will have a Taylor expansion (near to a) of the form

$$g(z) = b_0 + b_1(z-a) + b_2(z-a)^2 + \cdots$$

with $b_0 \neq 0$. Multiplying these together, we find that

$$\frac{g(z)}{f(z)} = \frac{a_0 b_0}{(z-a)^m} + \frac{a_0 b_1 + a_1 b_0}{(z-a)^{m-1}} + \cdots,$$

once more recognisable as a Laurent expansion at a pole of order m. ∎

---

[4] Once its removable singularity is removed.

Using the zeros–poles relationship we can extract a little more insight into singularities, provided that we clarify exactly what it means for a function (real or complex) to diverge to infinity:

9.1.8 Definition and notes

1. A real function $f : A \to \mathbb{R}$ *tends to infinity as x tends to a* (and we write $f(x) \to \infty$ as $x \to a$) if, for each positive K, we can find $\varepsilon_K > 0$ such that $f(x) > K$ for every x in $A \setminus \{a\}$ such that $|x - a| < \varepsilon_K$.

2. A complex function $f : A \to \mathbb{C}$ *tends to infinity as z tends to a* (and we write either $f(z) \to \infty$ or $|f(z)| \to \infty$ as $z \to a$) if, for each positive K, we can find $\varepsilon_K > 0$ such that $|f(z)| > K$ for every z in $A \setminus \{a\}$ such that $|z - a| < \varepsilon_K$.

3. Note (1): In both cases we are assuming that the domain includes points arbitrarily close to a.

4. Note (2): In both cases the defining conditions are equivalent to $\dfrac{1}{f} \to 0$ (as $x \to a$ or as $z \to a$) provided that f does not take the value zero (and, in case 1, provided that the values of f are actually positive).

9.1.9 Corollary   If f has a pole at a then, as $z \to a$, $|f(z)| \to \infty$.

Proof   If f has a pole of order m, then $1/f$ (once its singularity is removed) has a zero of order m, so continuity tells us that $\dfrac{1}{f(z)} \to 0$ as $z \to a$. Therefore also $\left| \dfrac{1}{f(z)} \right| \to 0$ as $z \to a$ and the result follows.   ∎

9.1.10 Remark   Interestingly, the converse of the last result is also true: if (as $z \to a$) $|f(z)| \to \infty$, then both $\dfrac{1}{|f(z)|}$ and $\dfrac{1}{f(z)}$ tend to zero, so $1/f$ has a limit (of zero) and 9.1.6 assures us that its singularity can be removed by defining it to take the value zero at a (but evidently nowhere else close to a). Now that $1/f$ has a zero of some order, 9.1.7 shows that f has a pole of that order.

So it is only in the case of a pole that $|f(z)| \to \infty$ as we approach the singularity.

Since 9.1.6 and 9.1.10 give us reasonably clear pictures of how a regular function must behave when we approach a singularity of one of the first two types (*removable* and *poles*) it would be pleasant to be able to outline the expected behaviour approaching the third type, the isolated *essential* singularities. Unfortunately, no simple description at such a point is valid, for the behaviour near such a danger point is always extremely chaotic. It can be shown that, if a is an essential singularity of f and $D(a, r) \setminus \{a\}$ is any punctured disc centred on a (no matter how small), then the range of f on that tiny punctured disc is the whole of the complex plane with, at most, one point missing! This almost unimaginably complicated behaviour has, fortunately, practically no consequences for what we are trying to do in this text. The interested reader could choose to search upon the terms *Casorati-Weierstrass theorem* or *Picard's great theorem* for additional insights.

9.2 The residue theorem

Let's first of all reinforce and recapitulate the definition that emerged in part (5) of paragraph 9.1.2:

9.2.1 Definition   If f is regular on (at least) a punctured disc $D \setminus \{a\} = \{z \in \mathbb{C} : 0 < |z - a| < r\}$ then its *residue at a* is the coefficient of $\dfrac{1}{z - a}$ in its Laurent expansion. We denote it by $\mathrm{Res}(f, a)$.

This is not an intrinsically attractive definition in most folks' opinion, but it does lean on how residues are usually calculated: one either finds the relevant part of the Laurent expansion in question, or uses an appropriate shortcut (see Section 9.3 for a selection of these) based on that expansion.

An alternative definition (which we get immediately from Laurent's theorem in Section 9.1) is:

9.2.2 Alternative definition   If f is regular on (at least) a punctured disc $D \setminus \{a\} = \{z \in \mathbb{C} : 0 < |z - a| < r\}$ then its *residue at a* is

$$\frac{1}{2\pi i} \int_K f(z)\, dz$$

in which K is any (anticlockwise) circular contour within the punctured disc and centred on a.

That, at first sight, is a scarier definition because it seems to suggest that, in order to determine a residue, one must take on the potentially difficult task of integrating the relevant function around a suitable circle. In reality, though, this second definition flags up what residues are really useful for: read it backwards, and what it tells us is that we can calculate the integral of a function around one of its singularities just by somehow spotting one coefficient in its Laurent expansion *and without doing any actual integrating at all*.

9.2.3 Example   Reviewing paragraphs 9.1.2 and 9.1.3 in this notation, we have

- $\mathrm{Res}\left(\dfrac{1 - \cos z}{z^5}, 0\right) = -\dfrac{1}{24}$,
- $\mathrm{Res}\left(e^{\frac{1}{z}}, 0\right) = 1$,

- Res $\left(\dfrac{z - \sin z}{z^n}, 0 \right) = 0$ if n is 3 or 5, $\frac{1}{6}$ if n is 4 and $-\frac{1}{120}$ if n is 6,

- Res $\left(\dfrac{2}{1 - \cos(2z)}, 0 \right) = 0$,

- Res $\left(\dfrac{1}{z^2 + 1}, i \right) = -\dfrac{i}{2}$ and the same method (although there are more efficient methods coming) will show that

- Res $\left(\dfrac{1}{z^2 + 1}, -i \right) = \dfrac{i}{2}$.

If you have thought about the 'geometry' of Cauchy's theorem and Cauchy's integral formula, you may well have realised that the fact that the contour around which we are presently integrating is a circle is very largely irrelevant: all that matters about it is that it is a simple closed contour with anti-clockwise orientation. Let's make that insight explicit:

9.2.4 Lemma   Suppose that f is regular in and on a simple closed anticlockwise contour Γ except for a single point a in the interior of Γ. Then $\int_\Gamma f(z)\,dz = 2\pi i \operatorname{Res}(f, a)$.

Proof

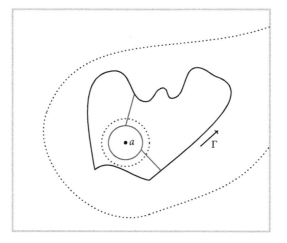

From 7.6.1 we have that

$$\int_\Gamma f(z)\,dz = \int_C f(z)\,dz$$

where C is a (suitably small) anticlockwise circular path centred on a. From 9.2.2 this is $2\pi i$ times the residue of f at a.   ■

The harder question—but it is only slightly harder—is this: how could we use these ideas to calculate integrals, around a contour, of a function that has *several* singularities inside the contour instead of just one as has been the case so far? The answer, though it counts as a major theorem, is as simple as you could reasonably wish for: we only need to count the contribution from each singularity in turn, and add them up.

9.2.5 **Cauchy's residue theorem**   Suppose that G is an open set that contains a simple closed anticlockwise contour Γ and its interior, and that the function f is regular on G except for a finite number of singularities (none of which is exactly on the track of Γ). Then

$$\int_\Gamma f(z)\,dz = 2\pi i \sum \operatorname{Res}(f, a_k),$$

the sum being taken over all the singularities that f has inside Γ.

Perhaps we ought to restate that in less formal language, in order to bring out how powerful it is: the integral of *almost any function* around *almost any such contour* is $2\pi i$ times the sum of the residues inside the contour. The only *caveats* are that the contour must not go through any singularity (which, of course, would have created confusion as to whether such a singularity was inside or outside) and that we only have a finite number of singularities to worry about.

(Sketch) proof

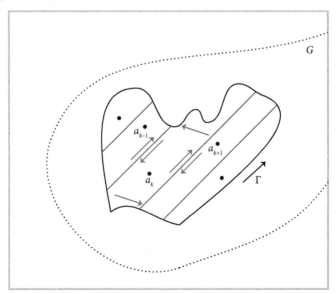

Cut up the region comprising the track plus the interior of Γ into slices, using straight line segments[5] so that each slice (S_k) contains just one singularity (a_k). If we integrate f around (the anticlockwise boundary of) S_k, the previous lemma gives us

$$\int_{S_k} f(z)\, dz = 2\pi i\, \mathrm{Res}(f, a_k).$$

Now add all these together. On the left-hand sides, the back-and-forth integrations along each cut will cancel out (because each construction cut is traversed twice but in opposite directions) and all the other integrations build up the integral around Γ. The right-hand sides total to $2\pi i$ times the sum of the residues within Γ, as required. ∎

9.3 Residue calculation tools

Paragraph 9.1.3 made the point that if we can easily identify the Laurent series (or its principal part) of a function at one of its singularities, then we can simply read off what its residue is. This is not always the quickest way to proceed, however, for there are a number of shortcuts for identifying residues that avoid having to delve into series expansions at all. Here are some of them:

9.3.1 Lemma   If f has a *simple* pole at a then

$$\mathrm{Res}(f, a) = \lim_{z \to a}(z - a)f(z).$$

Proof   Near to $z = a$ we have

$$f(z) = \frac{a_{-1}}{z - a} + a_0 + a_1(z - a) + a_2(z - a)^2 + \cdots$$

where a_{-1} is the required residue. Multiplying by $z - a$ and taking the limit as $z \to a$ immediately yields a_{-1}. ∎

9.3.2 Lemma   If f has a *double* pole at a then

$$\mathrm{Res}(f, a) = \lim_{z \to a}\left(\frac{d}{dz}\left((z - a)^2 f(z)\right)\right).$$

Proof   Near to $z = a$, this time we have

$$f(z) = \frac{a_{-2}}{(z - a)^2} + \frac{a_{-1}}{z - a} + a_0 + a_1(z - a) + a_2(z - a)^2 + \cdots$$

---

[5] For instance, choose a direction that is not parallel to any straight line from one singularity to another.

where a_{-1} is the required residue. Multiplying by $(z-a)^2$ and differentiating gets rid of the a_{-2} and leaves us with $a_{-1} + 2a_0(z-a) + \cdots$ whose limiting value as z approaches a is, indeed, a_{-1}. ∎

(Similar lemmas for residues at *triple* poles and higher can be derived, if needed, by the same kind of arguments.)

9.3.3 Example   Use 9.3.2 to find the residue at the origin of $\csc^2 z$ (without having to juggle power series as we did in 9.1.3 (8)).

Solution   Now $z^2 \csc^2 z = \dfrac{z^2}{\sin^2 z}$ differentiates via the quotient rule to give

$$\frac{2z\sin^2 z - 2z^2 \sin z \cos z}{\sin^4 z} = \frac{2z\sin z - 2z^2 \cos z}{\sin^3 z} = 2\left(\frac{z}{\sin z}\right)\left(\frac{\sin z - z\cos z}{\sin^2 z}\right).$$

The first bracket has a limit of 1 (as $z \to 0$). The second, using l'Hôpital, has the same limit as

$$\frac{z\sin z}{2\sin z\cos z} = \frac{z}{2\cos z}$$

which converges to zero. Therefore using 9.3.2 we have that the residue is zero.

One of the most useful such procedures is a spin-off from 9.3.1. It does not appear to have any generally accepted name in the literature, and for convenience we shall refer to it informally as the f/g' lemma. (The reason for the nickname will immediately become obvious.)

9.3.4 Lemma   Suppose that f and g are both regular on (at least) an open disc centre a, that g has a *simple* zero at a and[6] that $f(a) \neq 0$. Then the residue of $\dfrac{f(z)}{g(z)}$ at its singularity at a is given by

$$\operatorname{Res}\left(\frac{f(z)}{g(z)}, a\right) = \frac{f(a)}{g'(a)}.$$

Proof   The pole for $\dfrac{f(z)}{g(z)}$ is simple, as you can see from paragraph 9.1.7 (3). Then from 9.3.1 the residue is

$$\lim_{z\to a}(z-a)\frac{f(z)}{g(z)} = \lim_{z\to a}\frac{f(z)}{\left(\frac{g(z)}{z-a}\right)} = \lim_{z\to a}\frac{f(z)}{\left(\frac{g(z)-g(a)}{z-a}\right)} = \frac{f(a)}{g'(a)}.$$ ∎

---

[6] The result is also valid when $f(a) = 0$, but then it is not very interesting since $f(z)/g(z)$ has only a removable singularity at a.

9.3.5 Example   Evaluate the residue at $z = i$ of the function

$$\frac{e^{iz^3}}{(z^2 + 1)(z + 2i)^3}.$$

Solution   The top line $f(z) = e^{iz^3}$ is regular and non-zero everywhere in $\mathbb{C}$, and the bottom line $g(z) = (z^2 + 1)(z + 2i)^3$, also regular everywhere, has simple zeros at i and at $-i$, but a triple zero at $-2i$. By 9.1.7 this tells us that the given function $f(z)/g(z)$ has simple poles at i and at $-i$ together with a triple pole at $-2i$.
Using 9.3.4:

$$\text{Res}\left(\frac{f(z)}{g(z)}, i\right) = \frac{f(i)}{g'(i)} = \frac{e^{i(i)^3}}{3(i^2 + 1)(i + 2i)^2 + 2i(i + 2i)^3} = \frac{e^1}{0 + 2i(27i^3)} = \frac{e}{54}.$$

9.4 Exercises

1. Obtain the Laurent expansion, and identify the residue, of the following:
   (a) $\dfrac{z^4 + 2z^2 + 3}{z^2 + 1}$ at $z = i$,
   (b) $z^{-2}\sin^2(\pi z/4)$ at zero,
   (c) $z^3 e^{1/z}$ at zero.

2. For each of the following functions, identify the singularities, the type of each singularity and the residue at each singularity ((3) of paragraph 9.1.7 will often prove useful in such investigations):
   (a) $\dfrac{z}{z^2 + 100}$,
   (b) $\dfrac{e^z}{z^3(z^3 + 1)}$,
   (c) $\dfrac{1}{(z^2 + 1)^2(z^2 - 1)^2}$,
   (d) $z^{-3}e^{\sin z}$,
   (e) $\dfrac{\cos(z^3)}{z^6 - 1}$.

3. (a) Find the principal part of the Laurent expansion of (i) $\cot z$ at $z = 0$, (ii) $\tan z$ at $z = \frac{\pi}{2}$.

   (b) Find the principal part of the Laurent expansion of (i) $\cot^2 z$ at $z = 0$, (ii) $\tan^2 z$ at $z = \frac{\pi}{2}$.

4. For the function $f(z) = z^{-5}e^z \cos z$, find the principal part of the Laurent expansion at 0, and identify the residue there.

5. Obtain the principal part of the Laurent expansion at 0 of the function

$$f(z) = z^{-9}e^{-z^2}\cos(z^3).$$

Use this expansion to write down the value of each of the following integrals (around a small (anticlockwise) circle C centred on 0):

$$\int_C z^{-1} e^{-z^2} \cos(z^3)\, dz, \quad \int_C z^{-2} e^{-z^2} \cos(z^3)\, dz, \quad \int_C z^{-7} e^{-z^2} \cos(z^3)\, dz.$$

6. Evaluate the residues of the following functions at the singularities indicated:

(a) $\dfrac{e^{iz}}{z^2 + 4}$ at $z = 2i$,

(b) $\dfrac{\cos(\pi z)}{z^3 + z^2 + z + 1}$ at $z = -1$,

(c) $\dfrac{e^z \cos z}{z^6}$ at $z = 0$.

7. Carefully sketch the contour γ given by

$$\gamma(t) = \begin{cases} 2e^{2\pi i t} & \text{if } 0 \le t \le 2, \\ 1 + e^{-2\pi i t} & \text{if } 2 \le t \le 3. \end{cases}$$

Now evaluate

$$\int_\gamma \frac{e^{\sin(\pi z/2)}}{(z^2 - 1)(z^2 + 16)}\, dz.$$

(*Hint*: notice that γ can be conveniently expressed as $C_1 \oplus C_2 \ominus C_3$ where C_1, C_2 and C_3 are *simple anticlockwise* circular contours.)

8. For the function $f(z) = \dfrac{ze^{iz}}{z^4 + 5z^2 + 4}$ and the contours defined as

$$\gamma_1(t) = 3i + 3e^{it}, \quad 0 \le t \le 2\pi, \qquad \gamma_2(t) = 3i + 3e^{it}, \quad 0 \le t \le 6\pi,$$

$$\gamma_3(t) = 3i + 3e^{-it}, \quad 0 \le t \le 2\pi, \qquad \gamma_4(t) = 3i + 3e^{-it}, \quad 0 \le t \le 10\pi,$$

evaluate

$$\int_{\gamma_1} f(z)\, dz, \quad \int_{\gamma_2} f(z)\, dz, \quad \int_{\gamma_3} f(z)\, dz \text{ and } \int_{\gamma_4} f(z)\, dz.$$

9. (a) Obtain and prove a formula, based on 9.3.2, for calculating the residue of a function at a triple pole.

(b) Hence determine $\text{Res}\left(\dfrac{z^2}{(z^2 + 9)^3}, 3i\right)$.

10. The function $f(z) = \dfrac{1}{(z^2 - 9)(z^2 - 36)}$ has singularities at $-6, -3, 3$ and 6.
(Notice that the distances between $2 + i$ and these four points are all different.)
Investigate how the contour integral

$$\int_{C(2+i,r)} f(z)\, dz$$

varies with r, the radius of the (simple closed anticlockwise) circular contour indicated. (Exclude from consideration any values of r that cause the circle to pass directly through one of the singularities.)

11. (a) Given that f has only a finite number of singularities, and that we can find two positive constants R_0 and K such that $|f(z)| < \dfrac{K}{|z|^2}$ whenever $|z| > R_0$, use estimation of the integral to show that

$$\int_{C(0,R)} f(z)\, dz \to 0 \quad \text{as} \quad R \to \infty$$

(where the circular contour $C(0, R)$ is, as usual, closed, simple and anticlockwise).

(b) Consider a function f defined in the following form:

$$f(z) = \frac{z^n + a_{n-1}z^{n-1} + a_{n-2}z^{n-2} + \cdots + a_1 z + a_0}{z^{n+2} + b_{n+1}z^{n+1} + b_n z^n + \cdots + b_1 z + b_0}.$$

Use part (a) to show that, no matter how the positive integer n and the various coefficients a_k, b_k are chosen, the sum of the residues of f at all its singularities must be zero.

12. Suppose that f has a simple zero at a. Show that

$$\int_C \frac{zf'(z)}{f(z)}\, dz = 2\pi i a$$

for a sufficiently small (anticlockwise) circle C centred on a. (*Hint*: begin by using 8.4.2 to factorise f near to a.)

13. Let n be a positive integer and $\gamma(n)$ the (simple, closed, anticlockwise) rectangular contour having vertices $n + \frac{1}{2} + i, -n - \frac{1}{2} + i, -n - \frac{1}{2} - i, n + \frac{1}{2} - i$ (and back to $n + \frac{1}{2} + i$). Evaluate

$$\int_{\gamma(n)} \cot(\pi z)\, dz.$$

14. The function $f(z)$ has an isolated essential singularity at $z = a$. Prove that every disc in $\mathbb{C}$ includes a value of f (in the standard jargon, that the range of f is *dense* in $\mathbb{C}$).

Hint: use a proof by contradiction; if it were not so, then there would be some open disc $D(b, r)$ that contained *no* value of f; what does that tell us about the function

$$g(z) = \frac{1}{f(z) - b}$$

and, in consequence, about its behaviour as $z \to b$?

15. Continuing from the previous exercise, show that if f has an isolated essential singularity anywhere in $\mathbb{C}$, then *every* complex number is the limit of a sequence of values of f.

10 Reality from complexity

10.1 Integrating 'around the unit circle'

The connecting thread from here on is that it is often possible to solve a difficult or lengthy real integration problem by converting it into an easier or shorter complex integration question. This is so contrary to any reasonable expectation that could be built on the everyday meaning of the word 'complex' that, in order to render the benefit clearly visible, we need to take the considerable trouble of carrying out a few illustrative integrations doggedly within the field of real analysis.

Consider the problem of evaluating

$$\int_{c}^{d} \frac{1}{1 + a \cos \theta + b \sin \theta} \, d\theta$$

for various values of a, b, c and d. This question raises two kinds of difficulty, of which the first and most pressing is existential: does the integral even exist? To explore this at leisure, we can begin by plotting the graph of $a \cos \theta + b \sin \theta$ over a reasonably wide range of values of θ. What emerges is a classic sinusoidal wave of a certain amplitude (we shall see why this is in a moment):

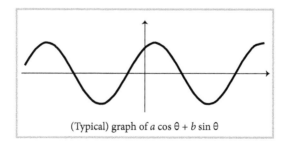

(Typical) graph of $a \cos \theta + b \sin \theta$

The graph of $1 + a \cos \theta + b \sin \theta$ can now be created by moving the previous graph one unit up the page; we illustrate this with a and/or b being 'large' compared with 1, and again with a and b being 'small' compared with 1:

Integration with Complex Numbers. McCluskey and McMaster, Oxford University Press.
© Brian McMaster and Aisling McCluskey (2022). DOI: 10.1093/oso/9780192846075.003.0010

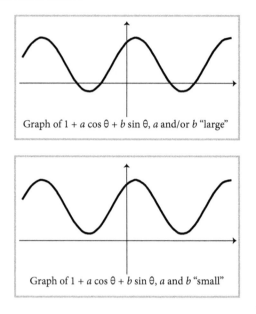

Graph of $1 + a \cos \theta + b \sin \theta$, a and/or b "large"

Graph of $1 + a \cos \theta + b \sin \theta$, a and b "small"

The second of these three diagrams shows why there is a potential existential issue here: it is possible that $1 + a \cos \theta + b \sin \theta$ takes the value zero somewhere in the domain of integration and, if this happens, then the integrand we set out to deal with is undefined at such points and the integral itself is improper (and therefore may be divergent).

Here is a worked example intended to alert us to the reality that this danger is not merely potential: even in quite simple cases the desired integral fails to exist.

10.1.1 **Example**   Investigate the integral

$$\int_0^\pi \frac{1}{1 + \cos \theta} \, d\theta.$$

Solution   Using the double-angle formula $\cos(2x) = 2 \cos^2 x - 1$ we can rewrite the indefinite integral here as

$$\int \frac{1}{2 \cos^2(\theta/2)} \, d\theta = \frac{1}{2} \int \sec^2(\theta/2) \, d\theta = \tan(\theta/2) + C$$

at which point the fundamental theorem of calculus wants to tell us that the definite integral is the change-in-value of that expression from $\theta = 0$ to $\theta = \pi$. Unfortunately, however, when $\theta = \pi$ the expression $\tan(\theta/2)$ is undefined—which forces our attention onto something that we should have realised earlier: the integrand $\frac{1}{1 + \cos \theta}$ is undefined at $\theta = \pi$, the original integral is for that reason improper, and what we should have been seeking to do is to evaluate

$$\int_0^{\pi-\varepsilon} \frac{1}{1+\cos\theta} \, d\theta$$

for small positive ε, and then looking for a limit as $\varepsilon \to 0$.

Now that the task has been clarified, the fundamental theorem can be invoked to get

$$\int_0^{\pi-\varepsilon} \frac{1}{1+\cos\theta} \, d\theta = [\tan(\theta/2)]_{\theta=0}^{\theta=\pi-\varepsilon} = \tan\left(\frac{\pi}{2} - \frac{\varepsilon}{2}\right).$$

However, as $\varepsilon \to 0$, the expression $\tan\left(\frac{\pi}{2} - \frac{\varepsilon}{2}\right)$ does not possess a limit (it tends to infinity). In other words, the integral we are exploring here is not just improper, it diverges (in effect, it does not exist as a real number).

In order to identify the circumstances in which this kind of improperness can arise, we need to recall a standard trigonometric rearrangement:

$$a\cos\theta + b\sin\theta = \sqrt{a^2+b^2}\left(\frac{a}{\sqrt{a^2+b^2}}\cos\theta + \frac{b}{\sqrt{a^2+b^2}}\sin\theta\right)$$

$$= \sqrt{a^2+b^2}\left(\sin\alpha\cos\theta + \cos\alpha\sin\theta\right) = \sqrt{a^2+b^2}\sin(\theta+\alpha)$$

where α is an angle chosen so that $\sin\alpha = \dfrac{a}{\sqrt{a^2+b^2}}$ and $\cos\alpha = \dfrac{b}{\sqrt{a^2+b^2}}$. The diagram

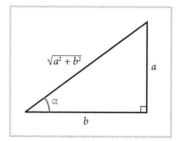

may be helpful in visualising this choice. With this in mind, the denominator of our original question can be reformulated as

$$1 + \sqrt{a^2+b^2}\sin(\theta+\alpha)$$

and from this we see that

- the range of values taken by $1 + \sqrt{a^2+b^2}\sin(\theta+\alpha)$ runs from $1 - \sqrt{a^2+b^2}$ to $1 + \sqrt{a^2+b^2}$,

- if $\sqrt{a^2 + b^2} < 1$ then $1 + a\cos\theta + b\sin\theta = 1 + \sqrt{a^2 + b^2}\sin(\theta + \alpha)$ can never take the value zero, but
- if $\sqrt{a^2 + b^2} \geq 1$ then $1 + a\cos\theta + b\sin\theta = 1 + \sqrt{a^2 + b^2}\sin(\theta + \alpha)$ does take the value zero for some values of θ (that can be identified by routine trigonometry).

In other words, the integral

$$\int_c^d \frac{1}{1 + a\cos\theta + b\sin\theta}\,d\theta$$

will always be proper (that is, not improper!) provided that $\sqrt{a^2 + b^2} < 1$, but will be improper for some values of c and d if $\sqrt{a^2 + b^2} \geq 1$.

From this point on, and for this reason, we handle only the case where $\sqrt{a^2 + b^2} < 1$, in other words, where $a^2 + b^2 < 1$.

Incidentally, the trigonometric rearrangement we used earlier can easily be tweaked to run as

$$a\cos\theta + b\sin\theta = \sqrt{a^2 + b^2}\left(\frac{a}{\sqrt{a^2 + b^2}}\cos\theta + \frac{b}{\sqrt{a^2 + b^2}}\sin\theta\right)$$

$$= \sqrt{a^2 + b^2}\,(\cos\beta\cos\theta + \sin\beta\sin\theta) = \sqrt{a^2 + b^2}\cos(\theta - \beta)$$

where β is an angle chosen so that $\cos\beta = \dfrac{a}{\sqrt{a^2 + b^2}}$ and $\sin\beta = \dfrac{b}{\sqrt{a^2 + b^2}}$.

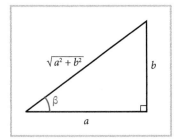

(The further consequence of these trig reorganisings is that it is often legitimate to assume that the denominator of these integration questions contains only '1+ a cosine term', or '1+ a sine term', since 'the other term' can be absorbed into the cosine or sine of a compound angle as shown above.)

Now that the existential crisis has been averted, we turn to the second kind of difficulty these integration questions raise—the technical, methodological kind. How does one actually evaluate one of these integrals in the context of real integration? The primary tool is, as usual, the fundamental theorem of calculus, but how in practice can we identify an expression whose derivative is $\dfrac{1}{1 + a\cos\theta + b\sin\theta}$?

10.1.2 Policy   For the indefinite integral

$$\int \frac{1}{1 + a\cos\theta + b\sin\theta}\, d\theta$$

implement the substitution $t = \tan(\theta/2)$.

10.1.3 Note   This rather un-obvious change of variable has the first benefit that all the trig ratios of θ can be expressed conveniently (but again un-obviously!) in terms of t. In particular, it is routine to confirm that

$$\sin\theta = \frac{2t}{1 + t^2} \quad \text{and} \quad \cos\theta = \frac{1 - t^2}{1 + t^2}$$

as the following diagram expresses visually:

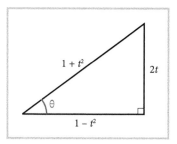

(We demonstrate the first of these identities thus:

$$\frac{2t}{1 + t^2} = \frac{2\tan(\theta/2)}{1 + \tan^2(\theta/2)} = \frac{2\sin(\theta/2)/\cos(\theta/2)}{\sec^2(\theta/2)}$$

$$= \frac{2\sin(\theta/2)\cos^2(\theta/2)}{\cos(\theta/2)} = 2\sin(\theta/2)\cos(\theta/2) = \sin\theta,$$

and the other is equally straightforward to check.)

The second benefit of the proposed substitution is how the '$d\theta$' is transformed into 'dt' in accordance with the standard change-of-variable procedures:

$$t = \tan(\theta/2) \Rightarrow \frac{dt}{d\theta} = \frac{1}{2}\sec^2(\theta/2) = \frac{1}{2}(1 + t^2)$$

so we must replace $d\theta$ by $\dfrac{2dt}{1 + t^2}$.

So the full effect of the substitution is

$$\int \frac{1}{1 + a\cos\theta + b\sin\theta}\, d\theta = \frac{1}{1 + \frac{a(1-t^2)}{1+t^2} + \frac{2bt}{1+t^2}} \frac{2dt}{1 + t^2}$$

$$= \int \frac{2}{1 + t^2 + a(1 - t^2) + 2bt} \, dt$$

$$= \int \frac{2}{(1 - a)t^2 + 2bt + (1 + a)} \, dt.$$

Although this is not what most people will call an *easy* integral, it is *doable* (at least, once you are given reasonable values of a and b). By way of illustration, consider the following:

10.1.4 **Example**   Evaluate

$$\int_{-\pi/2}^{\pi/2} \frac{1}{5 + 2 \cos \theta + 3 \sin \theta} \, d\theta.$$

Solution   Expressing this first as

$$\frac{1}{5} \int_{-\pi/2}^{\pi/2} \frac{1}{1 + 0.4 \cos \theta + 0.6 \sin \theta} \, d\theta$$

we see that the indefinite integral is the case $a = 0.4, b = 0.6$ of the discussion in paragraph 10.1.3, and therefore equals

$$\frac{1}{5} \int \frac{2}{0.6t^2 + 1.2t + 1.4} \, dt = \frac{1}{5} \int \frac{10}{3t^2 + 6t + 7} \, dt = \frac{2}{3} \int \frac{1}{\frac{4}{3} + (t + 1)^2} \, dt$$

(noting the device of 'completing the square' in the denominator) which is now a standard integral, giving

$$\frac{2}{3} \frac{\sqrt{3}}{2} \arctan \left(\frac{\sqrt{3}(t + 1)}{2} \right).$$

Lastly, the requested definite integral is the change in value of this expression from $\theta = -\pi/2$ to $\theta = \pi/2$, that is, from $t = \tan(-\pi/4) = -1$ to $t = \tan(\pi/4) = 1$. The answer now simplifies to

$$\frac{\sqrt{3}}{3} \left(\arctan \sqrt{3} - \arctan 0 \right) = \frac{\pi\sqrt{3}}{9}.$$

Two further methodological difficulties remain to be considered. If we alter the previous question and ask for

$$\int_0^{\pi} \frac{1}{5 + 2 \cos \theta + 3 \sin \theta} \, d\theta$$

then it seems natural to cannibalise the previous answer and simply write down

$$\left[\frac{\sqrt{3}}{3} \arctan \left(\frac{\sqrt{3}(\tan(\theta/2) + 1)}{2} \right) \right]_0^{\pi}.$$

However, substituting $\theta = \pi$ in this expression raises the spectre of $\tan(\pi/2)$ which is undefined. 'Schoolroom mathematics' suggests that we innocently replace $\tan(\pi/2)$ by ∞, argue that $\frac{\sqrt{3}(\infty + 1)}{2} = \infty$ whose inverse tangent is $\pi/2$, and come up with an overall answer of $\frac{\sqrt{3}}{3} \left(\frac{\pi}{2} - \arctan \left(\frac{\sqrt{3}}{2} \right) \right)$. This value is actually correct, but the logical invalidity of the argument should be obvious. A correct way to circumnavigate the hazard is to acknowledge the last display as improper, and treat it as

$$\lim_{\varepsilon \to 0^+} \left[\frac{\sqrt{3}}{3} \arctan \left(\frac{\sqrt{3}(\tan(\theta/2) + 1)}{2} \right) \right]_0^{\pi - \varepsilon}.$$

The cautionary point to be aware of is that, although $\theta = \pi$ was not a danger point (a point at which the definition failed) for the original integrand

$$\frac{1}{5 + 2 \cos \theta + 3 \sin \theta},$$

it became one for the transformed problem through the change of variable that we opted for.

To illustrate the second methodological difficulty that we want to flag up, consider how best to deal with

$$\int_0^{2\pi} \frac{1}{5 + 2 \cos \theta + 3 \sin \theta} \, d\theta.$$

An unthinking reliance on our earlier analysis might easily tempt us into writing down an 'answer' of

$$\left[\frac{\sqrt{3}}{3} \arctan \left(\frac{\sqrt{3}(\tan(\theta/2) + 1)}{2} \right) \right]_0^{2\pi}.$$

Yet this cannot possibly be right: the periodic nature of trig functions immediately evaluates it as zero, whereas even a rough sketch graph of the (positive!) function $\frac{1}{5 + 2 \cos \theta + 3 \sin \theta}$ over the interval $[0, 2\pi]$:

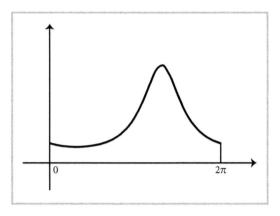

shows that the area between it and the horizontal axis is very definitely not zero. The underlying reason for failure here is that, as already pointed out, the point $\theta = \pi$ that falls in the middle of the domain of integration is one which renders the new variable $t = \tan(\theta/2)$ undefined, and so the reasons for trusting the change-of-variable result over the whole of $[0, 2\pi]$ break down. (It's an intriguing and salient point, that the all-too-common decision to ignore the *proof* of a theorem may lead to a perfectly understandable temptation to *use* that theorem in a situation where it does not apply and where it gives a plausible but quite incorrect answer!) To deal correctly with this integral, we should break it up into $\int_0^\pi$ and $\int_\pi^{2\pi}$, the first of which we have done already, and the second of which can be processed in very much the same way.

A final point we ought to make is that, even without being given 'numerically nice' values of a and b, it is perfectly possible to carry out the integration displayed at the end of 10.1.3 provided that $a^2 + b^2 < 1$; merely by taking enough care over the elementary algebra, we can convert

$$\int \frac{2}{(1-a)t^2 + 2bt + (1+a)}\, dt$$

into

$$\frac{2}{1-a} \int \frac{1}{\frac{1-a^2-b^2}{(1-a)^2} + \left(t + \frac{b}{1-a}\right)^2}\, dt$$

which yields

$$\frac{2}{\sqrt{1-a^2-b^2}} \arctan\left(\frac{(1-a)t + b}{\sqrt{1-a^2-b^2}}\right).$$

10.1.5 Summary   Provided always that $a^2 + b^2 < 1$, the substitution $t = \tan(\theta/2)$ will convert the indefinite integral

$$\int \frac{1}{1 + a\cos\theta + b\sin\theta}\, d\theta$$

into an integral with respect to t that can be carried out by elementary methods (and especially if we first combine the sine and cosine terms into a single trig function of a compound angle). However,

- the algebra and trigonometry and calculus, although elementary, are quite substantial and may be prone to casual error,
- if an odd multiple of π is an endpoint of the domain of a corresponding definite integral, then we should be aware that limiting processes are necessary in order correctly to evaluate it,
- if an odd multiple of π is an interior point of the domain of integration, then it is essential to split the domain at each such point and separately to evaluate each piece of the definite integral (unless, as sometimes happens, there is some symmetry in the question that can save us the bother). This issue in particular is easy to overlook, and especially damaging if we fail to pay attention to it.

10.1.6 Going complex Now let us revamp the entire question as an integral involving complex numbers. The (anticlockwise) unit circle C is conveniently described by

$$z = e^{i\theta}, \quad 0 \le \theta \le 2\pi$$

and in this notation we invoke Euler's formula, de Moivre's theorem and the fact that cosine is an even function whereas sine is an odd function to see that

$$z + \frac{1}{z} = 2\cos\theta, \quad z - \frac{1}{z} = 2i\sin\theta, \quad \text{so} \quad \cos\theta = \frac{1}{2}\left(z + \frac{1}{z}\right), \quad \sin\theta = \frac{1}{2i}\left(z - \frac{1}{z}\right).$$

Furthermore, $\dfrac{dz}{d\theta} = ie^{i\theta} = iz$ so the change-of-variable formula (swopping $d\theta$ for $\dfrac{dz}{iz}$) that defines integration around C gives us

$$\int_0^{2\pi} \frac{1}{1 + a\cos\theta + b\sin\theta}\, d\theta = \int_C \frac{1}{1 + \frac{a}{2}(z + z^{-1}) + \frac{b}{2i}(z - z^{-1})}\, \frac{dz}{iz}$$

$$= \frac{1}{i}\int_C \frac{2}{2z + a(z^2 + 1) - bi(z^2 - 1)}\, dz$$

$$= \frac{2}{i}\int_C \frac{1}{(a - bi)z^2 + 2z + (a + bi)}\, dz.$$

Now the integral here (according to Cauchy's residue theorem 9.2.5) is $2\pi i$ times the sum of the residues within C of the integrand. Given values of a and b this is pretty easy to compute: let us, for instance and comparison, rerun an earlier question:

10.1.7 Example   Evaluate

$$\int_0^{2\pi} \frac{1}{5 + 2\cos\theta + 3\sin\theta}\, d\theta.$$

Solution   This is (*one fifth of*) the special case $a = 0.4$, $b = 0.6$ so the transformed integral will be

$$\frac{2}{i}\int_C \frac{1}{(0.4 - 0.6i)z^2 + 2z + (0.4 + 0.6i)}\, dz = \frac{2}{i}\int_C \frac{5}{(2 - 3i)z^2 + 10z + (2 + 3i)}\, dz.$$

The quadratic formula readily tells us that the denominator has (simple) zeros at

$$\frac{-10 \pm \sqrt{100 - 4(13)}}{2(2 - 3i)} = \frac{-5 \pm 2\sqrt{3}}{2 - 3i}.$$

Now the zero with the minus sign on its square root evidently lies outside C, but the other (let us temporarily call it α) has modulus less than 1. Thus the sole residue within C of $\dfrac{5}{(2 - 3i)z^2 + 10z + (2 + 3i)}$ is $\dfrac{5}{2(2 - 3i)z + 10}$ evaluated at α (using the f/g' lemma, paragraph 9.3.4), that is, $\dfrac{5}{-10 + 4\sqrt{3} + 10} = \dfrac{5}{4\sqrt{3}}$. Gathering up the arithmetical fragments, our overall answer is

$$(2\pi i)\left(\frac{2}{i}\right)\left(\frac{1}{5}\right)\left(\frac{5}{4\sqrt{3}}\right) = \frac{\pi\sqrt{3}}{3}.$$

10.1.8 Notes

1. Most people, once they have got used to the little rituals of complex analysis, consider that the actual labour involved in 10.1.7 is considerably less than that in 10.1.4.

2. But perhaps the more important point is that, unlike the $t = \tan(\theta/2)$ procedure, the conversion of this kind of real integral into a complex integral around the unit circle does not introduce any risk of undefined quantities, does not create potential improper evaluations that ought to be implemented by limiting processes, and does not necessitate breaking up the domain of integration into 'safe' regions. The same condition $a^2 + b^2 < 1$ guarantees *both* that the original integral is not an improper integral, *and* that the transformed (complex) integrand has no singularities on the circle C.

3. On the other hand, the complex approach is useful only in the case where the limits of integration for θ are 2π apart (unless some symmetry consideration can be brought into play), for we cannot go 'only part-way around the circle' by this technique.

4. Once again, if we are sufficiently careful with the manipulations, we can deal with the general case of this family of integrals by integrating around C, as we now show.

10.1.9 Example   Evaluate, in the general case,

$$\int_0^{2\pi} \frac{1}{1 + a\cos\theta + b\sin\theta}\, d\theta$$

provided that $a^2 + b^2 < 1$.

Solution   By the earlier discussion, the integral equals

$$\frac{2}{i} \int_C \frac{1}{(a - bi)z^2 + 2z + (a + bi)}\, dz.$$

Via the quadratic formula (and temporarily writing $w = a + bi$) the denominator has (simple) zeros at $\dfrac{-2\pm\sqrt{4 - 4w\overline{w}}}{2\overline{w}} = \dfrac{-1\pm\sqrt{1 - |w|^2}}{\overline{w}}$.

Now $|\overline{w}| = \sqrt{a^2 + b^2}$ which we know to be less than 1, so $-1 - \sqrt{1 - |w|^2}$ certainly has modulus greater than $|\overline{w}|$, that is, $\dfrac{-1 - \sqrt{1 - |w|^2}}{\overline{w}}$ lies *outside* the unit circle. On the other hand, the product of the roots of this quadratic has to be $\dfrac{w}{\overline{w}}$ whose modulus is 1 so, since one root lies outside C (i.e. has modulus greater than 1) the other must lie inside (i.e. have modulus less than 1). We therefore know that the other root

$$\alpha = \frac{-1 + \sqrt{1 - |w|^2}}{\overline{w}}$$

is the unique singularity lying inside C. The residue at α is $\dfrac{1}{2(a - bi)z + 2}$ evaluated at α, that is, $\dfrac{1}{2\sqrt{1 - |w|^2}}$. Via the residue theorem, we find that the desired integral is

$$(2\pi i)\frac{2}{i}\left(\frac{1}{2\sqrt{1 - |w|^2}}\right) = \frac{2\pi}{\sqrt{1 - a^2 - b^2}}.$$

(Again we could have simplified the calculations somewhat by combining the cosine and sine terms into a compound sine or cosine before beginning... but, frankly, it wasn't necessary.)

We round off this section with a couple of worked examples at the level of difficulty and amount of detail appropriate to let the learner demonstrate the skills and techniques we intend to deliver.

10.1.10 Example   Use contour integration to evaluate

$$I = \int_0^{2\pi} \frac{1}{13 + 5 \cos \theta} \, d\theta.$$

Solution   The unit circle C is described by $z = e^{i\theta} = \cos\theta + i \sin\theta$ where $0 \le \theta \le 2\pi$. Then also $z + z^{-1} = 2 \cos \theta$ and $\dfrac{dz}{d\theta} = ie^{i\theta} = iz$. So

$$I = \int_C \frac{1}{13 + \frac{5}{2}(z + z^{-1})} \frac{dz}{iz} = \frac{2}{i} \int_C \frac{1}{5z^2 + 26z + 5} \, dz = \frac{2}{i} \int_C \frac{1}{(5z + 1)(z + 5)} \, dz.$$

The integrand here (let us call it $F(z)$ for short) has simple poles (simple zeros of the bottom line) at -5 and $-\frac{1}{5}$, and only the second one is inside C.

Using the f/g' shortcut, we get $\operatorname{Res}\left(F, -\frac{1}{5}\right) = \dfrac{1}{10z + 26}$ evaluated at $-\frac{1}{5}$, that is, $\frac{1}{24}$.

By the residue theorem,

$$I = 2\pi i \left(\frac{2}{i}\right) \frac{1}{24} = \frac{\pi}{6}.$$

10.1.11 Example   Use contour integration to evaluate

$$I = \int_0^{2\pi} \frac{1}{17 + 8 \cos \theta + 9 \sin \theta} \, d\theta.$$

Solution   Describe the unit circle C by $z = e^{i\theta} = \cos\theta + i \sin\theta$ with $0 \le \theta \le 2\pi$. Then also $z + z^{-1} = 2 \cos \theta, z - z^{-1} = 2i \sin \theta$ and $\dfrac{dz}{d\theta} = ie^{i\theta} = iz$. So

$$I = \int_C \frac{1}{17 + 4(z + z^{-1}) + \frac{9}{2i}(z - z^{-1})} \frac{dz}{iz}.$$

The integrand here tidies to

$$\frac{2}{(9 + 8i)z^2 + 34iz + (-9 + 8i)}$$

and, using the quadratic formula, the bottom line has simple zeros at $\alpha = \dfrac{-29i}{9 + 8i}$ and at $\beta = \dfrac{-5i}{9 + 8i}$. Now α has modulus greater than 1 and β does not, so β is the only (simple) pole of the integrand that lies inside C.

The residue of the integrand at β is

$$\frac{2}{2(9 + 8i)\beta + 34i} = \frac{1}{12i}.$$

Using lastly the residue theorem, we get

$$I = 2\pi i \frac{1}{12i} = \frac{\pi}{6}.$$

10.2 Integrating 'around an infinite semicircle'

Continuing the chapter's broad theme of 'calculating difficult real integrals by easy complex methods', we next turn to rational functions. A rational function (as we mentioned back in 3.1.2) is one that can be expressed as one polynomial divided by another (and the variable involved may be either real or complex). A typical rational function can therefore be written as $\dfrac{p(x)}{q(x)}$ or $\dfrac{p(z)}{q(z)}$ (where p and q are the polynomials in question) or, if it is necessary to spell out the full details, as

$$\frac{a_0 + a_1 x + a_2 x^2 + \cdots + a_n x^n}{b_0 + b_1 x + b_2 x^2 + \cdots \cdots + b_m x^m}$$

(but it usually isn't necessary).

As far as this section is concerned, we are going to assume that the coefficients are real numbers, and that the denominator does not have any real zeros—a reason for both decisions being that we are aiming to calculate real integrals of the form

$$\int_a^b \frac{p(x)}{q(x)} \, dx$$

over a large stretch of the real line, and if $q(x)$ were to take the value zero at some real number x, that would make the problem an improper integral (an additional difficulty which, at this stage, we can well do without).

We shall also work mainly with what we shall call *bottom-heavy* rationals, that is, those in which the degree of the denominator exceeds the degree of the numerator by at least 2 (in other words, in the last display but one above, those for which $m \geq n + 2$). It is not so easy to explain at this stage exactly why we want that restriction but, roughly speaking, it concerns the behaviour of the integrals when the domain of integration is allowed to stretch off towards infinity: for instance, $\frac{1}{x^2}$ is a (very simple) bottom-heavy rational whose integral $\int_1^K \frac{1}{x^2}\,dx$ is easily calculated as $1 - \frac{1}{K}$, and we see that this expression tends to 1 as K tends to infinity, which allows us to conclude that

$$\int_1^\infty \frac{1}{x^2}\,dx = 1.$$

In contrast, $\frac{x}{x^2+1}$ is also a simple rational function, although not bottom-heavy this time since the degree of the bottom line is only 1 greater than that of the top line, and again the integral $\int_1^K \frac{x}{x^2+1}\,dx$ is easy enough to determine: this time it is $\frac{1}{2}\log(K^2+1) - \frac{1}{2}\log(2)$. However, as K tends to infinity in this scenario, the expression increases without limit (it tends to infinity also) forcing us to the conclusion that

$$\int_1^\infty \frac{x}{x^2+1}\,dx \text{ diverges—that is, does not exist (as a real number).}$$

These two examples illustrate what one may reasonably expect in this area: that bottom-heavy rationals generally do have (convergent) integrals over 'infinite' stretches of the real line, and that non-bottom-heavy rationals generally do not. (Exercise 1 of Section 10.6 provides a more precise insight into the matter.)

With enough care and patience, any of these bottom-heavy, non-zero-on-the-real-line, real-coefficients rational functions can be integrated over any bounded interval on the real line *by elementary methods*. Once again, however, the labour of doing so can be quite considerable. Let us try a fairly typical example next.

10.2.1 Example Integrate over $[-2, 2]$ and, if possible, over $[0, \infty)$ and $(-\infty, \infty)$, the function

$$f(x) = \frac{8x^2 - 5x + 37}{(x^2+4)(x^2+9)}.$$

Solution We appeal to the theory of partial fractions to break our complicated rational into a sum of simpler rational pieces. Its prediction here is that

$$\frac{8x^2 - 5x + 37}{(x^2+4)(x^2+9)} = \frac{Ax+B}{x^2+4} + \frac{Cx+D}{x^2+9}$$

for some choice of constants A, B, C and D. It is straightforward (if a little tedious) to clear the fractions and compare coefficients and/or substitute in some convenient values for x in order to find these constants. We eventually discover that

$$\frac{8x^2 - 5x + 37}{(x^2+4)(x^2+9)} = \frac{1-x}{x^2+4} + \frac{x+7}{x^2+9}.$$

Wishing to 'make the top line be the derivative of the bottom line', we break this up as

$$\frac{1}{x^2 + 4} - \frac{1}{2} \left(\frac{2x}{x^2 + 4} \right) + \frac{1}{2} \left(\frac{2x}{x^2 + 9} \right) + \frac{7}{x^2 + 9}.$$

All the objects in this list can now be integrated by inspection, giving

$$\frac{1}{2} \arctan \left(\frac{x}{2} \right) - \frac{1}{2} \log(x^2 + 4) + \frac{1}{2} \log(x^2 + 9) + \frac{7}{3} \arctan \left(\frac{x}{3} \right)$$

$$= \frac{1}{2} \log \left(\frac{x^2 + 9}{x^2 + 4} \right) + \frac{1}{2} \arctan \left(\frac{x}{2} \right) + \frac{7}{3} \arctan \left(\frac{x}{3} \right).$$

At this point we can use the fundamental theorem of calculus to evaluate

$$\int_{-2}^{2} f(x)\, dx = \frac{\pi}{4} + \frac{14}{3} \arctan \left(\frac{2}{3} \right).$$

Likewise, for any positive K,

$$\int_{0}^{K} f(x)\, dx = \frac{1}{2} \log \left(\frac{K^2 + 9}{K^2 + 4} \right) + \frac{1}{2} \arctan \left(\frac{K}{2} \right) + \frac{7}{3} \arctan \left(\frac{K}{3} \right) - \frac{1}{2} \log \left(\frac{9}{4} \right)$$

and, noting the limiting behaviour as $K \to \infty$, we find that

$$\int_{0}^{\infty} f(x)\, dx = \frac{1}{2} \log 1 + \frac{1}{2} \left(\frac{\pi}{2} \right) + \frac{7}{3} \left(\frac{\pi}{2} \right) - \frac{1}{2} \log \left(\frac{9}{4} \right) = \frac{17\pi}{12} - \log 3 + \log 2.$$

For negative K, the same methods give

$$\int_{K}^{0} f(x)\, dx = -\frac{1}{2} \log \left(\frac{K^2 + 9}{K^2 + 4} \right) - \frac{1}{2} \arctan \left(\frac{K}{2} \right) - \frac{7}{3} \arctan \left(\frac{K}{3} \right) + \frac{1}{2} \log \left(\frac{9}{4} \right)$$

and from the limiting behaviour as $K \to -\infty$ we get

$$\int_{-\infty}^{0} f(x)\, dx = -\frac{1}{2} \log 1 - \frac{1}{2} \left(\frac{-\pi}{2} \right) - \frac{7}{3} \left(\frac{-\pi}{2} \right) + \frac{1}{2} \log \left(\frac{9}{4} \right) = \frac{17\pi}{12} + \log 3 - \log 2.$$

Combining the two totals shows finally that

$$\int_{-\infty}^{\infty} f(x)\, dx = \frac{17\pi}{6}.$$

In line with earlier comments, that was *elementary*, but one would hesitate to call it either *easy* or *elegant*. Let's take a look at how contour integration can help us to an alternative approach.

The complex rational function

$$f(z) = \frac{8z^2 - 5z + 37}{(z^2 + 4)(z^2 + 9)}$$

has singularities at the zeros of its denominator, that is, at $2i$, $-2i$, $3i$ and $-3i$. They are simple zeros so they create simple poles for f, which will make it easier to calculate residues at them.

Imagine the contour γ_K created by the segment $[-K, K]$ on the real axis, closed off by the semicircular arc S_K of radius K centred on 0 and lying in the upper half plane, K being some large positive number (certainly greater than 3).

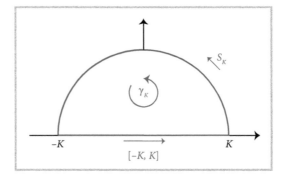

The two poles of f that are *inside* γ_K are $2i$ and $3i$. By the residue theorem,

$$\int_{\gamma_K} f = \int_{S_K} f + \int_{[-K,K]} f = 2\pi i (\text{Res}(f, 2i) + \text{Res}(f, 3i)).$$

Notice that the complex integral $\int_{[-K,K]} f(z)\, dz$ is just the real integral $\int_{-K}^{K} f(x)\, dx$.

Now *if only* we had a guarantee that, as $K \to \infty$, the integral $\int_{S_K} f$ around the enormous semicircle S_K tended to zero, this display would reduce the calculation of the CPV of $\int_{-\infty}^{\infty} f(x)\, dx$ to the simple arithmetic of calculating the two residues (and, of course, trying to remember to multiply their total by $2\pi i$).

Well, it turns out that we do indeed have such a guarantee:

10.2.2 Lemma   For any bottom-heavy rational $f(z) = \dfrac{p(z)}{q(z)}$, the integral $\int_{S_K} f(z)\, dz$ around the semicircular arc S_K tends to zero as $K \to \infty$.

Proof   Just this once, it helps to write out $f(z)$ in its full detail:

$$f(z) = \frac{a_0 + a_1 z + a_2 z^2 + \cdots + a_n z^n}{b_0 + b_1 z + b_2 z^2 + \cdots\cdots + b_m z^m}$$

where $m \geq n + 2$. Force-factorize that as $\dfrac{1}{z^2}$ times whatever it needs to be:

$$f(z) = \frac{1}{z^2} \left(\frac{a_0 z^2 + a_1 z^3 + a_2 z^4 + \cdots + a_n z^{n+2}}{b_0 + b_1 z + b_2 z^2 + \cdots\cdots + b_m z^m} \right).$$

The expression in the big parentheses tends to a limit as $z \to \infty$: the limit will be $\dfrac{a_n}{b_m}$ if m is exactly $n + 2$ (that is, if $f(z)$ was only just bottom-heavy) and it will be zero if m is bigger than that; but in both cases the limit exists, and therefore the bracketed expression is *bounded*: we can find a convenient constant M such that it is less than M in modulus provided that $|z|$ is big enough. Therefore, at every point on the semicircle S_K if K is big enough, we shall have $|f(z)| < M \left(\dfrac{1}{K^2} \right)$. So we can use the 'estimation of the integral' theorem (6.6.16) to see that, for large enough values of K,

$$\left| \int_{S_K} f(z)\, dz \right| < M \left(\frac{1}{K^2} \right) L(S_K) = M \left(\frac{1}{K^2} \right) (\pi K) = \frac{\pi M}{K}.$$

Since this last tends to zero as $K \to \infty$, it follows that the integral does so also. ∎

Returning now to the unfinished business immediately before Lemma 10.2.2, we have

$$\int_{-K}^{K} f(x)\, dx = 2\pi i (\text{Res}(f, 2i) + \text{Res}(f, 3i)) - \int_{S_K} f(z)\, dz$$

and allowing K to tend to infinity, we conclude that

$$\text{CPV} \int_{-\infty}^{\infty} f(x)\, dx = 2\pi i (\text{Res}(f, 2i) + \text{Res}(f, 3i))$$

which we can now routinely calculate (again using the f/g' trick) as

$$2\pi i \left(\frac{8z^2 - 5z + 37}{4z^3 + 26z} \Big|_{z=2i} + \frac{8z^2 - 5z + 37}{4z^3 - 26z} \Big|_{z=3i} \right) = 2\pi i \left(\frac{-17i}{12} \right) = \frac{17\pi}{6}.$$

Since that was a rather heavy example, and it was interrupted by its supporting lemma, we'll offer another of the same general type.

10.2.3 Example   Investigate the integral, from $-\infty$ to ∞, of the real function f given by

$$f(x) = \frac{40x^2 + 48x + 64}{x^4 + 64}.$$

Solution

1. Method one: real integration

   The usual first move in such questions is to factorise the denominator and, although a sum of squares won't factorise in real numbers, a sum of fourth powers will: in the present case we find that

   $$x^4 + 64 = (x^2 - 4x + 8)(x^2 + 4x + 8).$$

   Now appeal to the theory of partial fractions to break our complicated rational into a sum of simpler rational pieces. Its prediction here is that

   $$\frac{40x^2 + 48x + 64}{x^4 + 64} = \frac{40x^2 + 48x + 64}{(x^2 - 4x + 8)(x^2 + 4x + 8)} = \frac{Ax + B}{x^2 - 4x + 8} + \frac{Cx + D}{x^2 + 4x + 8}$$

   for some choice of constants A, B, C and D. It is straightforward (but tedious) to identify these. We eventually discover that

   $$\frac{40x^2 + 48x + 64}{x^4 + 4} = \frac{4x + 10}{x^2 - 4x + 8} - \frac{4x + 2}{x^2 + 4x + 8}.$$

   Wishing to 'make the top line be the derivative of the bottom line', we rearrange this as

   $$= \frac{2(2x - 4) + 18}{x^2 - 4x + 8} - \frac{2(2x + 4) - 6}{x^2 + 4x + 8}.$$

   Next, where fractional expressions have only constants on the top line, we complete the square on the bottom line to push the expression towards a standard arctan integration:

   $$= \frac{2(2x - 4)}{x^2 - 4x + 8} + \frac{18}{(x - 2)^2 + 4} - \frac{2(2x + 4)}{x^2 + 4x + 8} + \frac{6}{(x + 2)^2 + 4}.$$

   All the objects in the list can now be integrated by inspection, giving:

   $$2\log(x^2 - 4x + 8) + 18\left(\frac{1}{2}\right)\arctan\left(\frac{x - 2}{2}\right) - 2\log(x^2 + 4x + 8)$$
   $$+ 6\left(\frac{1}{2}\right)\arctan\left(\frac{x + 2}{2}\right)$$

   $$= 2\log\left(\frac{x^2 - 4x + 8}{x^2 + 4x + 8}\right) + 9\arctan\left(\frac{x - 2}{2}\right) + 3\arctan\left(\frac{x + 2}{2}\right).$$

   We are now in a position to carry out definite integration over virtually any interval that may be asked. In particular:

- $$\int_0^K f(x)\,dx$$

$$= 2\log\left(\frac{K^2-4K+8}{K^2+4K+8}\right) + 9\arctan\left(\frac{K-2}{2}\right)$$

$$+ 3\arctan\left(\frac{K+2}{2}\right) - 2\log 1 - 9\arctan(-1) - 3\arctan(1)$$

$$= 2\log\left(\frac{K^2-4K+8}{K^2+4K+8}\right) + 9\arctan\left(\frac{K-2}{2}\right) + 3\arctan\left(\frac{K+2}{2}\right) + \frac{3\pi}{2}$$

which, as $K \to \infty$, converges to

$$\int_0^\infty f(x)\,dx = 2\log 1 + 9\left(\frac{\pi}{2}\right) + 3\left(\frac{\pi}{2}\right) + \frac{3\pi}{2} = \frac{15\pi}{2},$$

- $$\int_{-K}^0 f(x)\,dx = 2\log 1 + 9\arctan(-1)$$

$$+ 3\arctan(1) - 2\log\left(\frac{K^2+4K+8}{K^2-4K+8}\right) - 9\arctan\left(\frac{-K-2}{2}\right)$$

$$- 3\arctan\left(\frac{-K+2}{2}\right)$$

$$= -\frac{3\pi}{2} + 2\log\left(\frac{K^2-4K+8}{K^2+4K+8}\right) + 9\arctan\left(\frac{K+2}{2}\right) + 3\arctan\left(\frac{K-2}{2}\right)$$

which, as $-K$ tends to minus infinity, that is, as $K \to \infty$, tends to

$$\int_{-\infty}^0 f(x)\,dx = -\frac{3\pi}{2} + 2\log 1 + 9\left(\frac{\pi}{2}\right) + 3\left(\frac{\pi}{2}\right) = \frac{9\pi}{2}.$$

- Finally, $\int_{-\infty}^\infty f(x)\,dx = \int_{-\infty}^0 + \int_0^\infty = 12\pi.$

2. Method two: complex contour integration
   The complex rational function

$$f(z) = \frac{40z^2 + 48z + 64}{z^4 + 64}$$

has singularities at the zeros of its denominator, that is, where $z^4 = -64$, that is, at the four (complex) fourth roots of -64. These have modulus $2\sqrt{2}$ and argument angles $\pi/4, 3\pi/4, 5\pi/4$ and $7\pi/4$. Possibly more conveniently for arithmetical purposes, they are $2 + 2i, -2 + 2i, -2 - 2i$ and $2 - 2i$. They are simple zeros (for the fundamental theorem of algebra says that $z^4 + 64$ can only have four zeros counted by multiplicity, and we have already found four distinct ones) so they create simple poles for f, which renders the calculation of residues easy.

Imagine the contour γ_K created by the segment $[-K, K]$ on the real axis, closed off by the semicircular arc S_K of radius K centred on 0 and lying in the upper half plane, K being some large positive number (certainly greater than $2\sqrt{2}$).

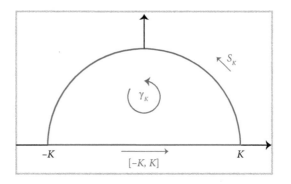

The two poles of f that are *inside* γ_K are $\alpha = 2 + 2i$ and $\beta = -2 + 2i$. By the residue theorem,

$$\int_{\gamma_K} f = \int_{S_K} f + \int_{[-K,K]} f = 2\pi i(\mathrm{Res}(f, \alpha) + \mathrm{Res}(f, \beta)).$$

Notice that the complex integral $\int_{[-K,K]} f(z)\, dz$ is just the real integral $\int_{-K}^{K} f(x)\, dx$. Lemma 10.2.2 assures us that, as $K \to \infty$, the integral $\int_{S_K} f$ around the semicircle S_K tends to zero. So now we have

$$\int_{-K}^{K} f(x)\, dx = 2\pi i(\mathrm{Res}(f, \alpha) + \mathrm{Res}(f, \beta)) - \int_{S_K} f(z)\, dz$$

and allowing K to tend to infinity, we conclude that

$$CPV \int_{-\infty}^{\infty} f(x)\, dx = 2\pi i (\mathrm{Res}(f, \alpha) + \mathrm{Res}(f, \beta))$$

which we can now routinely calculate using the f/g' algorithm from 9.3.4, yielding

$$2\pi i \left(\frac{40z^2 + 48z + 64}{4z^3} \Big|_{z=\alpha} + \frac{40z^2 + 48z + 64}{4z^3} \Big|_{z=\beta} \right) = 2\pi i(-6i) = 12\pi.$$

The conscientious reader will be concerned that, while using contour integration in 10.2.3, we have only determined the Cauchy principal value of the integral: there might still be a possibility that $\int_{-\infty}^{0} f(x)\, dx$ and $\int_{0}^{\infty} f(x)\, dx$ both diverge 'in opposite directions' in such a way that their combined approximation $\int_{-K}^{K} f(x)\, dx$ converges to a limit. However, pretty routine algebra will verify that

$$|f(x)| = \left| \frac{40x^2 + 48x + 64}{x^4 + 64} \right| \le \frac{M}{x^2}$$

for large values (positive or negative) of x, M being a constant (any value of M greater than 40 will work because, as x tends to $\pm\infty$, $f(x)$ behaves like $\frac{40x^2}{x^4}$). Once that is understood, a use of 5.5.6 will quickly confirm that the integral does converge, and therefore the Cauchy principal value 'gets it right' (See also paragraph 5.6.6.)

An additional benefit of the complex contour approach to these real integration questions is now becoming apparent. Quite apart from whether it is generally simpler or shorter than the standard real approach, it tends to be more uniform and less *ad hoc*. The next group of examples illustrates this even more emphatically.

10.2.4 **Example**   Investigate the integrals from 0 to ∞ and from $-\infty$ to ∞ of:

1. $\dfrac{1}{x^6 + 1}$,

2. $\dfrac{x}{x^6 + 1}$,

3. $\dfrac{x^2}{x^6 + 1}$,

4. $\dfrac{x^3}{x^6 + 1}$,

5. $\dfrac{x^4}{x^6 + 1}$,

6. $\dfrac{x^5}{x^6 + 1}$.

Partial real-integration solution   Two of these six can be done so easily through *real* calculus methods that there is no reason to consider using complex techniques on them. Firstly, in number (6) the substitution $u = 1 + x^6$ gives us that

$$\int \frac{x^5}{1 + x^6} \, dx = \frac{1}{6} \log(1 + x^6) \; (+ \text{ an arbitrary constant})$$

so the definite integral from 0 to K is $\frac{1}{6} \log(1 + K^6)$ which diverges as $K \to \infty$, and the integral from 0 to ∞ does not exist as a real number. (Similar remarks apply to the integral from $-\infty$ to 0.) Then in number (3) the substitution $u = x^3$ yields

$$\int \frac{x^2}{1 + x^6} \, dx = \frac{1}{3} \arctan x^3 \; (+ \text{ an arbitrary constant})$$

and so this time the definite integral from 0 to K is $\frac{1}{3} \arctan K^3$ which converges to $\frac{\pi}{6}$ as $K \to \infty$. Considerations of symmetry show that

$$\int_{-\infty}^{0} \frac{x^2}{1 + x^6} \, dx = \frac{\pi}{6} \text{ and } \int_{-\infty}^{\infty} \frac{x^2}{1 + x^6} \, dx = \frac{\pi}{3}.$$

Item (2) can be evaluated by combining substitution and partial fractions. Begin by recalling that $1 + u^3$ factorises as $(1 + u)(1 - u + u^2)$. Now using the substitution $u = x^2$ we see that

$$\int \frac{x}{1 + x^6} \, dx = \frac{1}{2} \int \frac{du}{1 + u^3} = \frac{1}{2} \int \frac{du}{(1 + u)(1 - u + u^2)}$$

where the integrand breaks up into partial fractions, giving us

$$\frac{1}{6} \int \left(\frac{1}{1 + u} + \frac{2 - u}{1 - u + u^2} \right) du.$$

This integrates to yield $\frac{1}{6} \log(1 + u) - \frac{1}{12} \log(1 - u + u^2) + \frac{\sqrt{3}}{6} \arctan \left(\frac{2u - 1}{\sqrt{3}} \right)$, with u meaning x^2, from which we can read off the integral from 0 to ∞ as $\frac{\pi \sqrt{3}}{9}$. Symmetry (this being an odd function) deduces that

$$\int_{-\infty}^{\infty} \frac{x}{1 + x^6} \, dx = 0.$$

Item (4) uses very similar partial fractions and the same substitution $u = x^2$ to kick it off: we (eventually) get

$$\int \frac{x^3}{1 + x^6} \, dx = -\frac{1}{6} \log(1 + u) + \frac{1}{12} \log(1 - u + u^2) + \frac{\sqrt{3}}{6} \arctan \left(\frac{2u - 1}{\sqrt{3}} \right)$$

curiously like item (2), and resulting in exactly the same answers as item (2) had:

$$\int_0^\infty \frac{x^3}{1+x^6}\, dx = \frac{\pi\sqrt{3}}{9}, \quad \int_{-\infty}^\infty \frac{x^3}{1+x^6}\, dx = 0.$$

Items (1) and (5) appear to be genuinely technically harder to crack. Anyone with *enough* time and patience (or, better still, access to a good computer algebra package) can *verify* quite routinely that $\int 1/(x^6+1)\, dx =$

$$\frac{1}{12}\left(4\arctan x + 2\arctan(\sqrt{3}+2x) - 2\arctan(\sqrt{3}-2x)\right.$$
$$\left. + \sqrt{3}\log\left(\frac{1+\sqrt{3}x+x^2}{1-\sqrt{3}x+x^2}\right)\right)$$

and that $\int x^4/(x^6+1)\, dx =$ the remarkably similar

$$\frac{1}{12}\left(4\arctan x + 2\arctan(\sqrt{3}+2x) - 2\arctan(\sqrt{3}-2x)\right.$$
$$\left. + \sqrt{3}\log\left(\frac{1-\sqrt{3}x+x^2}{1+\sqrt{3}x+x^2}\right)\right)$$

but if there is a straightforward method of *discovering* these outcomes through the usual channels, then we have not come across it.

An important point is that, in seeking to determine these integrals by *real* calculus methods, the appropriate technology changes from question to question, and one needs to be alert to different substitutions, to the use or non-use of partial fractions, and to small but intricate algebraic reorganisations.

Complex-integration solution   Turning now to the complex approach, the 'down' side is that integration around the 'infinite semicircle' will not let us directly access integrals from zero to infinity and will only give Cauchy principal values for integrals from minus infinity to infinity, but the very substantial 'up' side is that exactly the same technique is appropriate for all such problems (excluding item (6) which, of course, is not bottom-heavy) and, furthermore, very little algebraic reorganising is called for. In fact, let us directly compute the CPV of

$$\int_{-\infty}^\infty \frac{x^n}{1+x^6}\, dx$$

for all integer values of n from 0 to 4 at once!

The denominator of the corresponding complex rational function $\dfrac{z^n}{1+z^6}$ has simple zeros in the upper half plane at $r_1 = e^{\pi i/6}$, $r_2 = i$ and $r_3 = e^{5\pi i/6}$. The relevant residue at the typical r_j is $\dfrac{z^n}{6z^5} = \frac{1}{6}z^{n-5}$ evaluated at r_j (via f/g' again—see 9.3.4) and in each case the integral around the so-called infinite semicircle is zero. So the answer is

$$CPV \int_{-\infty}^{\infty} \frac{x^n}{1 + x^6} \, dx = 2\pi i \left(\frac{1}{6}\right) \left(r_1^{n-5} + r_2^{n-5} + r_3^{n-5}\right)$$

which requires nothing more sophisticated than complex number arithmetic (which, indeed, a decent modern calculator will do for you). We get the following evaluations for the (Cauchy principal values of the) integrals from $-\infty$ to ∞:

$n=0$, integral $= \frac{2\pi}{3}$,

$n=1$, integral $= 0$,

$n=2$, integral $= \frac{\pi}{3}$,

$n=3$, integral $= 0$,

$n=4$, integral $= \frac{2\pi}{3}$

from which we can immediately deduce for $n = 0, 2$ and 4 (because the integrands here are *even* functions) that the integrals from zero to infinity are $\frac{\pi}{3}$, $\frac{\pi}{6}$ and $\frac{\pi}{3}$ respectively.

10.2.5 Exercise   Evaluate the integrals from 0 to ∞, from $-\infty$ to 0, and from $-\infty$ to ∞ of

$$f(x) = \frac{x^2}{x^4 - 4x^2 + 16}.$$

Suggestion   We get the first answer to be $\frac{\pi}{4}$.

10.2.6 Exercise   How many of the integrals from 0 to ∞ of

$$f(x) = \frac{x^n}{x^8 + a^8}$$

(where a is a positive real constant and $0 \le n \le 6$) can you evaluate without an excessive amount of work?

10.2.7 Exercise   Evaluate

$$\int_{-\infty}^{\infty} \frac{1}{(1 + x^4)(1 + x^2)} \, dx, \quad \int_{-\infty}^{\infty} \frac{x^2}{(1 + x^4)(1 + x^2)} \, dx \text{ and}$$

$$\int_{-\infty}^{\infty} \frac{x^4}{(1 + x^4)(1 + x^2)} \, dx.$$

(The answers should be $\pi/2$, $\pi(\sqrt{2} - 1)/2$ and $\pi/2$ respectively.)
Explain why we can easily deduce the integrals from 0 to ∞ of the same three functions.

10.2.8 Exercise Explore how

$$\int_0^\infty \frac{x^2}{(1+x^4)(a^2+x^2)}\, dx$$

varies with the positive parameter a.

10.3 Spiking it with trig

The integrals

$$\int_{-\infty}^\infty \frac{\sin(2x)}{x^4+81}\, dx, \quad \int_{-\infty}^\infty \frac{\cos(5x)}{(x^2+1)(x^2+36)}\, dx$$

appear, at first sight, to be very different from those we studied in Section 10.2. Certainly the theory of (real) partial fractions will not help us to break up these integrands into simpler pieces—it applies only to pure rational functions. And if we do the obvious thing, converting them into complex functions merely by replacing x by z thus:

$$\int \frac{\sin(2z)}{z^4+81}\, dz, \quad \int \frac{\cos(5z)}{(z^2+1)(z^2+36)}\, dz$$

it doesn't help a great deal: in particular because the complex sine and cosine are more complicated functions than their real counterparts. (For instance, the useful inequalities $|\sin x| \le 1$, $|\cos x| \le 1$, which simplify so many estimates in real calculus, do not carry over to the complex case: the complex sine and cosine are both unbounded.)

However, if instead we recall Euler's formula and perceive $\cos\theta$ and $\sin\theta$ as the real and imaginary parts of $e^{i\theta}$, something much more encouraging begins to unfold. For one thing, e^{iaz} does not have any poles or zeros, so the modified integrals

$$\int \frac{e^{2iz}}{z^4+81}\, dz, \quad \int \frac{e^{5iz}}{(z^2+1)(z^2+36)}\, dz$$

experience no more (and no fewer) residues than were already present in their associated rational functions. More encouraging (and less obvious) is the fact that Lemma 10.2.2, concerning the disappearance of integrals of certain types of function along the infinite semicircle, is still perfectly valid for this sort of function also:

10.3.1 Lemma For any bottom-heavy rational $f(z) = \dfrac{p(z)}{q(z)}$ and any positive real constant a, the integral $\int_{S_K} e^{iaz} f(z)\, dz$ around the semicircular arc S_K tends to zero as $K \to \infty$.

Proof   Recall that when we write out $f(z)$ in full detail:

$$f(z) = \frac{a_0 + a_1 z + a_2 z^2 + \cdots + a_n z^n}{b_0 + b_1 z + b_2 z^2 + \cdots \cdots + b_m z^m}$$

we can reshape it as

$$f(z) = \frac{1}{z^2} r(z)$$

where $r(z)$ tends to a limit as $|z| \to \infty$, and is consequently *bounded*: we can find a constant M such that $|r(z)| \le M$ provided that $|z|$ is big enough.

So, what effect does the extra factor e^{iaz} have on the integral? Expressing z as $x + iy$ we see that

$$\left| e^{iaz} \right| = \left| e^{iax - ay} \right| = \left| e^{iax} \right| \left| e^{-ay} \right| = e^{-ay}.$$

Yet at every point of the semicircle S_K (within the upper half plane $y \ge 0$, of course, and remembering that a is positive) we get $e^{-ay} \le e^0 = 1$! Therefore the integral estimation that we did back in 10.2.2 hardly changes at all:

$$\left| \int_{S_K} e^{iaz} f(z)\, dz \right| < (1)(M)\left(\frac{1}{K^2} \right) L(S_K) = M\left(\frac{1}{K^2} \right)(\pi K) = \frac{\pi M}{K}$$

which tends to zero as $K \to \infty$. Hence, so does the (smaller) integral, as claimed. ∎

The effect of this lemma is to allow us to evaluate (at least the Cauchy principal values of) integrals such as the two we began this section with, almost exactly as if the trigonometric 'spikes' weren't there: their only impact being to alter the values of the residues. So let's now use it on some integrals of this form:

10.3.2 **Example**   Find the Cauchy principal value of

$$\int_{-\infty}^{\infty} \frac{\cos(5x)}{x^2 + 4}\, dx.$$

Solution   The complex function $\dfrac{e^{5iz}}{z^2 + 4}$ has, in the upper half plane, just the one simple pole at $2i$. With $K > 2$ and γ_K composed of the real interval $[-K, K]$ and the semicircular arc S_K in the upper half plane, we get from the residue theorem that

$$\int_{[-K,K]} \frac{e^{5iz}}{z^2+4} \, dz + \int_{S_K} \frac{e^{5iz}}{z^2+4} \, dz = \int_{\gamma_K} \frac{e^{5iz}}{z^2+4} \, dz = 2\pi i \operatorname{Res}(f, 2i)$$

so, taking limits as $K \to \infty$ (and using our new lemma 10.3.1),

$$\mathrm{CPV} \int_{-\infty}^{\infty} \frac{\cos(5x) + i\sin(5x)}{x^2+4} \, dx = 2\pi i \operatorname{Res}(f, 2i).$$

The relevant residue is $\dfrac{e^{5iz}}{2z}$ evaluated at $2i$, which calculates easily as $\dfrac{e^{-10}}{4i}$.
Thus we conclude that

$$\mathrm{CPV} \int_{-\infty}^{\infty} \frac{\cos(5x) + i\sin(5x)}{x^2+4} \, dx = \frac{\pi}{2e^{10}}$$

and, equating real and imaginary parts,

$$\mathrm{CPV} \int_{-\infty}^{\infty} \frac{\cos(5x)}{x^2+4} \, dx = \frac{\pi}{2e^{10}}, \quad \mathrm{CPV} \int_{-\infty}^{\infty} \frac{\sin(5x)}{x^2+4} \, dx = 0 \text{ (of course).}$$

Once again, although we have only determined the Cauchy principal values, each of the real functions under scrutiny is smaller in modulus than $\dfrac{1}{1+x^2}$, and it is easy to confirm that the improper integral

$$\int_{-\infty}^{\infty} \frac{1}{1+x^2} \, dx$$

is convergent. Using the absolute convergence and direct comparison notions we flagged up in 5.5.5 and 5.5.6, it follows that these integrals are also convergent, and that we have determined not just their Cauchy principal values but their actual values.

10.3.3 **Example**   Find the (Cauchy principal) value of

$$\int_{-\infty}^{\infty} \frac{\sin(2x + \theta)}{x^4 + 81} \, dx$$

for fixed $\theta \in \mathbb{R}$.

Solution   We shall firstly investigate the integrals of the slightly simpler functions $\sin(2x)/(x^4+81)$ and $\cos(2x)/(x^4+81)$. The complex function $\dfrac{e^{2iz}}{z^4+81}$ has, in the upper half plane, simple poles at the relevant fourth roots of -81, that is, at $\alpha = 3e^{\pi i/4}$, $\beta = 3e^{3\pi i/4}$. With $K > 3$ and γ_K composed of the real interval $[-K, K]$ and the semicircular arc S_K in the upper half plane, we get from the residue theorem that

$$\int_{[-K,K]} \frac{e^{2iz}}{z^4 + 81} \, dz + \int_{S_K} \frac{e^{2iz}}{z^4 + 81} \, dz = \int_{\gamma_K} \frac{e^{2iz}}{z^4 + 81} \, dz = 2\pi i (\text{Res}(f, \alpha) + \text{Res}(f, \beta))$$

so, taking limits as $K \to \infty$ (and using Lemma 10.3.1),

$$\text{CPV} \int_{-\infty}^{\infty} \frac{\cos(2x) + i\sin(2x)}{x^4 + 81} \, dx = 2\pi i (\text{Res}(f, \alpha) + \text{Res}(f, \beta)).$$

Then the relevant residues are $\dfrac{e^{2iz}}{4z^3}$ evaluated at α and at β, whose sum calculates out as

$$\frac{-i(\cos(3\sqrt{2}) + \sin(3\sqrt{2}))}{54\sqrt{2}e^{3\sqrt{2}}}.$$

(Do use a decent calculator for that bit of drudgery!)

Thus we conclude that

$$\text{CPV} \int_{-\infty}^{\infty} \frac{\cos(2x) + i\sin(2x)}{x^4 + 81} \, dx = \frac{\pi\sqrt{2}}{54e^{3\sqrt{2}}} (\cos(3\sqrt{2}) + \sin(3\sqrt{2}))$$

and, equating real and imaginary parts,

$$\text{CPV} \int_{-\infty}^{\infty} \frac{\cos(2x)}{x^4 + 81} \, dx = \frac{\pi\sqrt{2}}{54e^{3\sqrt{2}}} (\cos(3\sqrt{2}) + \sin(3\sqrt{2})),$$

$$\text{CPV} \int_{-\infty}^{\infty} \frac{\sin(2x)}{x^4 + 81} \, dx = 0.$$

(Again, the second part of the conclusion should have been obvious some time ago since the integrand is an odd function.)

Lastly, since $\sin(2x + \theta) = \sin(2x)\cos\theta + \cos(2x)\sin\theta$, the required (Cauchy principal) value is

$$\frac{\pi\sqrt{2}\sin\theta}{54e^{3\sqrt{2}}} (\cos(3\sqrt{2}) + \sin(3\sqrt{2})).$$

10.3.4 Exercise   Modify this argument to find the (Cauchy principal) value of

$$\int_{-\infty}^{\infty} \frac{\cos(nx)}{x^2 + a^2} \, dx$$

for arbitrary positive integer n and positive real a.

10.3.5 Example   Find the (Cauchy principal) value of

$$\int_{-\infty}^{\infty} \frac{\cos x}{(x^2 + 1)(x^2 + 4)} \, dx.$$

Solution   The complex function $\dfrac{e^{iz}}{(z^2 + 1)(z^2 + 4)}$ has, in the upper half plane, simple poles at i and at $2i$. With $K > 2$ and γ_K composed of the real interval $[-K, K]$ and the usual semicircular arc S_K, we get from the residue theorem that

$$\int_{[-K,K]} \frac{e^{iz}}{(z^2 + 1)(z^2 + 4)} \, dz + \int_{S_K} \frac{e^{iz}}{(z^2 + 1)(z^2 + 4)} \, dz = \int_{\gamma_K} \frac{e^{iz}}{(z^2 + 1)(z^2 + 4)} \, dz$$

$= 2\pi i$(sum of residues) so, taking limits as $K \to \infty$ (and using Lemma 10.3.1),

$$\mathrm{CPV} \int_{-\infty}^{\infty} \frac{\cos x + i \sin x}{(x^2 + 1)(x^2 + 4)} \, dx = 2\pi i(\text{sum of residues}).$$

These residues are $\dfrac{e^{iz}}{4z^3 + 10z}$ evaluated at i and at $2i$, which add up as $\dfrac{e^{-1}}{6i} + \dfrac{e^{-2}}{12i}$.

Thus

$$\mathrm{CPV} \int_{-\infty}^{\infty} \frac{\cos x + i \sin x}{(x^2 + 1)(x^2 + 4)} \, dx = \frac{\pi(2e - 1)}{6e^2}$$

so

$$\mathrm{CPV} \int_{-\infty}^{\infty} \frac{\cos x}{(x^2 + 1)(x^2 + 4)} \, dx = \frac{\pi(2e - 1)}{6e^2}.$$

10.3.6 Exercise   Revamp the solution of 10.3.5 to find the (Cauchy principal) value of

$$\int_{-\infty}^{\infty} \frac{\cos(nx)}{(x^2 + a^2)(x^2 + b^2)} \, dx$$

for any positive integer n and any *distinct* positive reals a and b. Also investigate the case $a = b$ *after* you have worked through the next paragraph and Exercise 10.3.7.

Be aware that, when a pole is not simple, a different technique for evaluating the residue there will become necessary: the f/g' shortcut (see 9.3.4) that has been our mainstay only works for *simple* poles, caused by *simple* zeros of the denominator. For double poles we can use paragraph 9.3.2 instead.

10.3.7 Exercise   Evaluate (the CPV of) the integral from $-\infty$ to ∞ of the functions

$$f(x) = \frac{1}{(x^2 + 16)^2}, \quad g(x) = \frac{\cos x}{(x^2 + 16)^2}.$$

Partial solution   Since the corresponding complex function $f(z)$ has, in the upper half plane, a *double* pole at $4i$, we invoke 9.3.2 to determine its residue there (which should turn out to be $\frac{1}{256i}$).

The rest of the argument should run exactly as before, leading to CPV $\int_{-\infty}^{\infty} f(x)\, dx = \frac{\pi}{128}$. It should also be routine to confirm that the integral actually converges.

Very much the same remarks apply to the function g. This time the residue should work out as $\frac{5e^{-4}}{256i}$ and the second integral as $\frac{5\pi}{128e^4}$.

10.4 Some special case techniques

The unit circle and the so-called infinite semicircle are only two out of an enormous range of simple closed contours that have proved useful in evaluating real integrals via complex techniques: indeed, there are so many 'special cases' that all we can reasonably do is to illustrate a handful of them. The common thread is, once you have decided on a complex function that on the real axis will give you the real function that you want, you need to select a contour along whose edges the integrals will be either very small or directly computable or closely related to the 'real' integral that is being sought *and, in particular, a contour that avoids the function's singularities.* We start with a variation on the 'bottom-heavy rationals' of Sections 10.2 and 10.3 where, as it turns out, the previous choice of contour is still powerful enough to deliver results.

VARIATION 1: NOT SO BOTTOM-HEAVY BUT EXPONENTIALISED

Although we can't integrate (from minus infinity to plus infinity) expressions like $\frac{x}{x^2 + a^2}$ by the method of Section 10.3 since it's not bottom-heavy enough, surprisingly we can do it for functions of the form $\frac{xe^{iax}}{x^2 + a^2}$. All we need is a couple of fairly straightforward results attributed to Jordan:

10.4.1 Jordan's inequality For every number θ such that $0 \le \theta \le \frac{\pi}{2}$, we have

$$\frac{2\theta}{\pi} \le \sin\theta \le \theta.$$

Remark You can think of this result graphically, as saying that (for acute angles) the graph of the sine function lies between the straight line '$y = x$' and the straight line joining the origin to the first maximum point on the sine graph, which makes it look very plausible

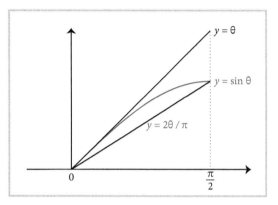

but which does not, of course, count as a logically watertight proof!

Proof of Jordan's inequality For $\theta = 0$ it is certainly true, so now we can suppose $0 < \theta \le \frac{\pi}{2}$ and consider θ as fixed for the moment. The function $\cos(\theta t)$ of t is easy to integrate while t runs from 0 to 1:

$$\int_0^1 \cos(\theta t)\, dt = \left[\frac{1}{\theta}\sin(\theta t)\right]_0^1 = \frac{\sin\theta}{\theta}.$$

Also, since the cosine function is decreasing on the interval $[0, \frac{\pi}{2}]$, we have (for each t between 0 and 1)

$$1 \ge \cos(\theta t) \ge \cos\left(\frac{\pi}{2}t\right),$$

and therefore

$$\int_0^1 1\, dt \ge \int_0^1 \cos(\theta t)\, dt \ge \int_0^1 \cos\left(\frac{\pi}{2}t\right) dt,$$

that is,

$$1 \ge \frac{\sin\theta}{\theta} \ge \frac{2}{\pi}\left[\sin\left(\frac{\pi}{2}t\right)\right]_0^1 = \frac{2}{\pi}.$$

Now multiplying across by (positive) θ gives the claimed result. ∎

Suppose now that S_R denotes (as usual) the semicircular arc described by

$$S_R(t) = Re^{it}, \quad 0 \le t \le \pi$$

for each positive radius R,

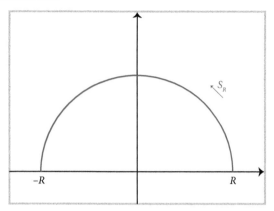

and that f is a complex function that is continuous on (at least) a set of the form

$$\{Re^{i\theta} : 0 \le \theta \le \pi, \ R > R_0\},$$

that is, on 'all points of the upper half plane that are far enough away from the origin'.

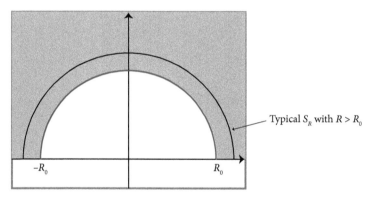

Typical S_R with $R > R_0$

This will already be enough to guarantee that $|f|$ (being also continuous where f was) will have a biggest value on each of the arcs S_R for which $R > R_0$, that is, the number $\max_{|z|=R} |f(z)|$ will exist. In this notation:

10.4.2 Jordan's lemma   Suppose that $\max_{|z|=R} |f(z)| \to 0$ as $R \to \infty$ (where, as in 10.4.1, f is continuous on the upper half plane sufficiently far from the origin). Then for each positive constant k, we have

$$\int_{S_R} e^{ikz} f(z)\, dz \to 0 \quad \text{as } R \to \infty.$$

This result is very close in spirit and in appearance to 10.2.2 which allowed us effectively to ignore the integrals of bottom-heavy rational functions on huge semicircles. The differences, both of which we might have expected to create difficulties for us (but they don't) are that 10.4.2 does not start with bottom-heaviness in full strength, but merely with $|f|$ being small on huge semicircles, and that 10.4.2 has a potentially complicating exponential factor. The proof is really just the standard estimation-of-integral trick *plus* Jordan's inequality *plus* the theorem about the modulus of an integral never exceeding the integral of the modulus (see paragraph 6.2.8).

Proof (of Jordan's lemma)   Let ε be any given positive number. Given that $\max_{|z|=R} |f(z)|$ tends to zero as R increases, we can find $R_1 > R_0$ such that $|f(z)| < \varepsilon$ whenever z lies on S_R and $R \geq R_1$. It follows from the definition of contour integral that (when $R \geq R_1$)

$$\left| \int_{S_R} e^{ikz} f(z)\, dz \right| = \left| \int_0^\pi e^{ikR(\cos\theta + i\sin\theta)} f(Re^{i\theta}) iRe^{i\theta}\, d\theta \right|$$

$$\leq R \int_0^\pi e^{-kR\sin\theta} |f(Re^{i\theta})|\, d\theta$$

$$\leq R\varepsilon \int_0^\pi e^{-kR\sin\theta}\, d\theta$$

$$= 2R\varepsilon \int_0^{\pi/2} e^{-kR\sin\theta}\, d\theta$$

(because the sine graph is symmetrical about the half-way point in the interval $[0, \pi]$)

$$\leq 2R\varepsilon \int_0^{\pi/2} e^{-2kR\theta/\pi}\, d\theta$$

(this is where we use Jordan's inequality *plus* the fact that exp is a monotone function. And now we have a function that is actually easy to integrate...)

$$= \frac{2R\varepsilon\pi}{2kR} \left(1 - e^{-kr} \right)$$

$$< \frac{\pi\varepsilon}{k}.$$

Since ε can be made arbitrarily small, this establishes that the integral does indeed tend to zero. ∎

10.4.3 **Example**   Evaluate

$$\int_0^\infty \frac{x \sin x}{x^2 + 1}\, dx.$$

Solution   Notice for a start that $x \sin x$ is the imaginary part of xe^{ix}. So we look at the function $h(z) = e^{iz} f(z)$ where

$$f(z) = \frac{z}{z^2 + 1}$$

which is regular in the upper half plane except for a simple pole at i, and (provided that $R > 1$) on the semicircular arc S_R we have $|z^2 + 1| \geq R^2 - 1$ which gives us

$$\left| \frac{z}{z^2 + 1} \right| \leq \frac{R}{R^2 - 1}$$

which tends to zero as $R \to \infty$. All the conditions of Jordan's lemma are therefore satisfied, and we are guaranteed that $\int_{S_R} h(z)\, dz \to 0$ (as $R \to \infty$). With γ_R denoting the contour comprising S_R and its bounding diameter $[-R, R]$, the residue theorem tells us

$$2\pi i \operatorname{Res}(h, i) = \int_{\gamma_R} h(z)\, dz = \int_{S_R} h(z)\, dz + \int_{[-R, R]} h(z)\, dz$$

$$= \int_{S_R} h(z)\, dz + \int_{-R}^R \frac{x(\cos x + i \sin x)}{x^2 + 1}\, dx$$

$$= \int_{S_R} h(z)\, dz + 2 \int_0^R \frac{x(i \sin x)}{x^2 + 1}\, dx$$

(notice that $x \cos x/(x^2 + 1)$ is an odd function, whereas $x \sin x/(x^2 + 1)$ is an even function).

Now we calculate the residue using the f/g' lemma, and we find that $\operatorname{Res}(h, i) = \dfrac{ie^{-1}}{2i} = \dfrac{1}{2e}$. The previous display now says

$$\frac{\pi i}{e} = \int_{S_R} h(z)\, dz + 2i \int_0^R \frac{x \sin x}{x^2 + 1}\, dx,$$

that is,

$$\int_0^R \frac{x \sin x}{x^2 + 1}\, dx = \frac{\pi}{2e} - \frac{1}{2i} \int_{S_R} h.$$

Taking limits as $R \to \infty$ yields

$$\int_0^\infty \frac{x \sin x}{x^2 + 1} \, dx = \frac{\pi}{2e}.$$

10.4.4 **Example**   Evaluate

$$\int_0^\infty \frac{x^3 \sin(\pi x)}{x^4 + 20x^2 + 64} \, dx.$$

Solution   Look at the function $h(z) = e^{\pi i z} f(z)$ given by

$$f(z) = \frac{z^3}{z^4 + 20z^2 + 64} = \frac{z^3}{(z^2 + 4)(z^2 + 16)}$$

which is regular in the upper half plane except for simple poles at $2i$ and $4i$. Provided always that $R > 4$, we have on the semicircular arc S_R

$$|(z^2 + 4)(z^2 + 16)| \geq (R^2 - 4)(R^2 - 16)$$

which gives us

$$\left| \frac{z^3}{z^4 + 20z^2 + 64} \right| \leq \frac{R^3}{(R^2 - 4)(R^2 - 16)}$$

which tends to zero as $R \to \infty$. All the conditions of Jordan's lemma are satisfied, so we deduce that $\int_{S_R} h(z) \, dz \to 0$ $(R \to \infty)$. With γ_R denoting the contour comprising S_R and its bounding diameter $[-R, R]$, the residue theorem now says

$$2\pi i (\text{Res}(h, 2i) + \text{Res}(h, 4i)) = \int_{\gamma_R} h(z) \, dz = \int_{S_R} h(z) \, dz + \int_{[-R,R]} h(z) \, dz$$

$$= \int_{S_R} h(z) \, dz + \int_{-R}^R h(x) \, dx$$

$$= \int_{S_R} h + 2i \int_0^R \frac{x^3 \sin(\pi x)}{x^4 + 20x^2 + 64} \, dx$$

(taking account of odd and even functions here).

We use 9.3.4 to calculate the two residues: $\text{Res}(h, 2i) = \dfrac{(2i)^3 e^{-2\pi}}{4(2i)^3 + 40(2i)} = $

$-\dfrac{e^{-2\pi}}{6}$ and $\text{Res}(h, 4i) = \dfrac{(4i)^3 e^{-4\pi}}{4(4i)^3 + 40(4i)} = \dfrac{2e^{-4\pi}}{3}$. From the previous display we now get

$$\int_0^R \frac{x^3 \sin(\pi x)}{x^4 + 20x^2 + 64} dx = \pi \left(\frac{2e^{-4\pi}}{3} - \frac{e^{-2\pi}}{6} \right) - \frac{1}{2i} \int_{S_R} h$$

and, taking limits as $R \to \infty$,

$$\int_0^\infty \frac{x^3 \sin(\pi x)}{x^4 + 20x^2 + 64} dx = \pi \left(\frac{2e^{-4\pi}}{3} - \frac{e^{-2\pi}}{6} \right) = \frac{\pi(4 - e^{2\pi})}{6e^{4\pi}}.$$

VARIATION 2: SKIRTING AROUND THE DIFFICULTY

So with our present techniques we can't handle things like $\frac{x}{x^2 - a^2}$ or $\frac{xe^{iax}}{x^2 - a^2}$ due to the bottom line having zeros actually on the real axis. However, it is sometimes possible almost literally to bypass such singularities: instead of integrating around a big semicircle as usual, you kink the diameter by a couple of *tiny* semicircular detours that avoid by epsilon the zeros on the real axis. Perhaps surprisingly, it works (provided that we can accept Cauchy principal values rather than full-blooded convergence).

To get some intuitive picture of what happens here, let's consider a particular question: to evaluate the Cauchy principal value of

$$\int_0^\infty \frac{1}{1 - x^4} dx.$$

To begin with, since the integrand is an even function, this (if we can evaluate it) will be exactly half of

$$\int_{-\infty}^\infty \frac{1}{1 - x^4} dx$$

so, at first sight, integrating around an infinite semicircle seems an appropriate method. Unfortunately, the natural complex version of the integrand

$$\frac{1}{1 - z^4}$$

has singularities at $1, i, -1$ and $-i$, and two of them lie exactly on the bounding diameter of the big semicircle that our previous experience leads us to think of. Since it is not possible to view these as being truly interior nor truly exterior to the semicircular contour, the residue theorem fails to apply. However, if we were to modify the contour using tiny semicircular detours either to include 1 and -1

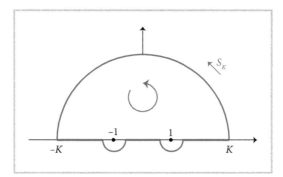

or to exclude them

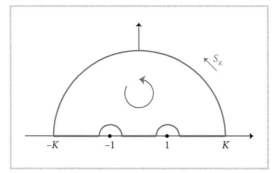

then the residue theorem would become applicable. The additional question that arises at this point is: how do we determine the integral along those tiny semi-circular indentations? A rational initial guess is that, since the integral around a complete (anticlockwise) circle containing a single one of the singularities is $2\pi i$ times the residue there, each semicircle might contribute half of that, that is, πi times the appropriate residue. It turns out—as we shall show—that the reality is not quite as simple as that, but not far off; to be precise, we shall find that:

10.4.5 **Proposition**   In the notation of Variation 2:

1. the integral along such a semicircle is, in some circumstances, *nothing at all like* half of the integral around the complete circle, but

2. if the singularity is a *simple pole*, then in the limit (as the radius of the indentation tends to zero) the semicircular integral is indeed half of the circular integral (that is, πi times the residue).

Proof

1. One of the easiest ways in which to establish our first point here is to consider the function $f(z) = z^{-2}$ and its unique singularity at 0. Since z^{-2} is 'its own Laurent series' we see that the residue—the coefficient of z^{-1}—is zero. Therefore

the integral of f around any circle centred on 0 is precisely zero. However, if we evaluate its integral along a typical semicircular arc centred on 0 (please see the next diagram)

$$\int_S f(z)\, dz, \quad S(t) = Re^{i(\alpha+t)}, \quad 0 \le t \le \pi$$

then, using the definition of path integral, we get

$$\int_0^\pi R^{-2} e^{-2i(\alpha+t)} Rie^{i(\alpha+t)}\, dt = i \int_0^\pi R^{-1} e^{-i(\alpha+t)}\, dt$$

$$= -R^{-1} \left[e^{-i(\alpha+t)} \right]_0^\pi = 2R^{-1} e^{-i\alpha}$$

which, as you can see, is completely unrelated to 'half of zero'. (For instance, when $\alpha = 0$ it is $2/R$, when $\alpha = \pi$ it is $-2/R$, and when $\alpha = \pi/2$ it is $-2i/R$. We point out in passing that all of these will actually diverge to infinity as the radius of the indentation shrinks towards 0.)

2. In contrast, suppose now that f has a simple pole at z_0. Fix an angle α and refer again to the next diagram. For small values of (radius) $R > 0$ consider the semicircle S_R described by

$$S_R(t) = z_0 + Re^{i(\alpha+t)}, \quad 0 \le t \le \pi$$

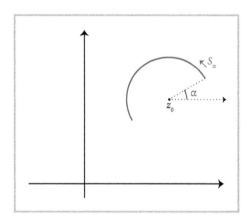

and let K denote the residue of f at z_0. Then for values of z reasonably close to z_0 we shall have

$$f(z) = g(z) + \frac{K}{z - z_0}$$

where the regular function g is simply the Taylor part of the Laurent series (and $\dfrac{K}{z - z_0}$ is the principal part). If we choose a closed disc $\overline{D}(z_0, \delta)$ on which this equation holds and let M, say, denote the biggest value of $|g(z)|$ on that closed disc, then for $0 < R < \delta$ we find that

$$\left| \int_{S_R} g \right| \leq ML(S_R) \to 0 \text{ as } R \to 0.$$

Therefore the limiting value of $\int f$ around S_R equals the limiting value of $\int \dfrac{K}{z - z_0}\, dz$ around S_R, which is easily calculated as $\pi i K$ (and is actually the same for all values of α and of R), as we claimed. ■

Now let's go back to our interrupted investigation:

10.4.6 Example   Find the Cauchy principal value of

$$\int_0^\infty \frac{1}{1 - x^4}\, dx.$$

Solution   If we choose a large value of R and a small value of ε, and consider the simple closed contour Γ comprising the semicircle (in the upper half plane and centred on zero) of radius R and its bounding diameter $[-R, R]$ indented by semi-circular diversions (also in the upper half plane) of radius ε centred at 1 and at -1,

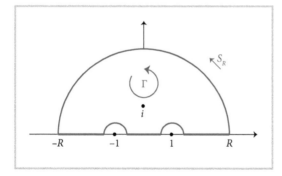

then the only singularity possessed by the function $f(x) = \dfrac{1}{1 - z^4}$ that lies interior to Γ is i, and the residue is $1/(-4z^3)$ evaluated at i, that is, $-i/4$. By the residue theorem, $\int_\Gamma f = 2\pi i(-i/4) = \pi/2$. Letting $R \to \infty$ and $\varepsilon \to 0$, we see as before that the integral around the 'big' semicircle tends to zero and, from part (2) of 10.4.5, each of the integrals along the 'small' semicircles will have limiting values of $-\pi i$ times the residue at the centre (and note that these semicircles are *clockwise* so the sign of the integral is switched). It follows that the limit of

$$\int_{[-R,-1-\varepsilon]\cup[-1+\varepsilon,1-\varepsilon]\cup[1+\varepsilon,R]} f(z)\, dz$$

is $\pi/2 + \pi i(\mathrm{Res}(f, -1) + \mathrm{Res}(f, 1)) = \pi/2 + \pi i\,(1/4 + (-1/4)) = \pi/2$ and that therefore, *in a limited sense,*

$$\int_0^\infty \frac{1}{1-x^4}\, dx = \frac{\pi}{4}.$$

It must be stressed that all we have determined here is the *Cauchy principal value* of each of the integrals $\int_{-\infty}^\infty f$ and $\int_0^\infty f$ because, in fact, both of these will be seen to diverge if we study them a little more carefully. For instance, provided that $0 \le x < 1$, straightforward calculus/real-analysis techniques will show that

$$\frac{1}{1-x^4} = \frac{1}{2}\frac{1}{1+x^2} + \frac{1}{4}\frac{1}{1+x} + \frac{1}{4}\frac{1}{1-x}$$

and

$$\int \frac{1}{1-x^4}\, dx = \frac{1}{2}\arctan x + \frac{1}{4}\log\frac{1+x}{1-x}$$

so that, for small positive ε,

$$\int_0^{1-\varepsilon} \frac{1}{1-x^4}\, dx = \frac{1}{2}\arctan(1-\varepsilon) + \frac{1}{4}\log\frac{2-\varepsilon}{\varepsilon}$$

which manifestly diverges to infinity as ε shrinks towards 0.

VARIATION 3: THINKING INSIDE THE BOX

Occasionally it pays dividends to integrate not around a big semicircle but around a long rectangle whose length tends to infinity. Of course you need to gather evidence that the integrals along three of the sides tend to zero (or are controllable in some other manner), and this can be a lengthy exercise needing both care and patience.

10.4.7 **Example (part 1)**   Given a real number a such that $0 < a < 1$, use the substitution $x = e^t$ to show that

$$\int_0^\infty \frac{x^{a-1}}{x+1}\, dx, \quad \int_{-\infty}^\infty \frac{e^{at}}{e^t+1}\, dt$$

are equivalent in the sense that, if either exists,[7] then so does the other, and they are then equal.

---

[7] In point of fact it is not difficult to prove that they do converge: see Exercise 7 in Chapter 5.

Solution   For any (large) positive R and (small) positive ε, when we substitute $x = e^t$ (and therefore $dx/dt = e^t$) in the integral

$$\int_\varepsilon^R \frac{x^{a-1}\, dx}{x + 1},$$

it turns into

$$\int_{\log \varepsilon}^{\log R} \frac{e^{at-t}e^t\, dt}{e^t + 1} = \int_{\log \varepsilon}^{\log R} \frac{e^{at}\, dt}{e^t + 1}.$$

Now letting $\varepsilon \to 0$ is the same as letting $\log \varepsilon \to -\infty$, and letting $R \to \infty$ is the same as letting $\log R \to \infty$. From this we can see the asserted equivalence.

10.4.8 Example (part 2)   By integrating the function $f(z) = e^{az}/(e^z + 1)$ around the rectangle B whose vertices in anticlockwise order are $R, R + 2\pi i, -S + 2\pi i, -S$ (and then back to R), investigate the integral

$$\int_{-\infty}^{\infty} f(x)\, dx$$

using Cauchy principal value if necessary.

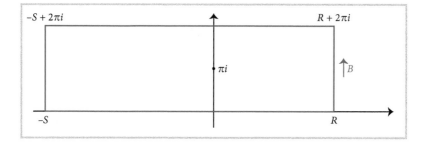

Solution   The integrand has a simple pole at πi but no others in or on this rectangle, and its residue at πi is $\dfrac{e^{\pi i a}}{e^{\pi i}} = -e^{\pi i a}$, so the residue theorem tells us that the integral is $-2\pi i e^{\pi i a}$. Notice that this is independent of both R and S.

On the rectangle's vertical right-hand side joining R to $R + 2\pi i$, described as a segment by setting $\gamma(t) = R + ti, \;\; 0 \le t \le 2\pi$, the integral is

$$I(right, R) = \int_0^{2\pi} \frac{e^{aR+ait}}{e^{R+it} + 1}i\, dt$$

whose modulus is at most

$$\int_0^{2\pi} \frac{e^{aR}}{e^R - 1} \, dt = \frac{2\pi e^{aR}}{e^R - 1}$$

which tends to zero as $R \to \infty$. So the integral $I(right, R)$ also tends to zero as $R \to \infty$.

By the same argument, the integral $I(left, S)$ along the opposite vertical segment tends to zero as $S \to \infty$.

The integral—let us denote it by $I(R, S)$—along the horizontal base of B is just the real integral

$$I(R, S) = \int_{-S}^R \frac{e^{ax}}{e^x + 1} \, dx.$$

Describing the upper horizontal edge (transcribed from right to left, of course!) by, for example, $\gamma(t) = -t + 2\pi i$, $-R \le t \le S$, we find that the relevant integral is

$$\int_{-R}^S \frac{e^{a(-t+2\pi i)}}{e^{-t+2\pi i} + 1} \, (-dt)$$

(substitute $x = -t$)

$$= \int_R^{-S} \frac{e^{a(x+2\pi i)}}{e^{x+2\pi i} + 1} \, dx$$

$$= e^{2\pi i a} \int_R^{-S} \frac{e^{ax}}{e^x + 1} \, dx = -e^{2\pi i a} \int_{-S}^R \frac{e^{ax}}{e^x + 1} \, dx = -e^{2\pi i a} I(R, S).$$

Putting the four pieces together we see that the integral around the box B, which we know to equal $-2\pi i e^{\pi i a}$, is given by $(1 - e^{2\pi i a})I(R, S) + I(right, R) + I(left, S)$. Rearranging the algebra gives us

$$I(R, S) = \frac{-2\pi i e^{\pi i a} - I(right, R) - I(left, S)}{1 - e^{2\pi i a}}.$$

If we now let R and S tend to infinity independently of one another (so this is not a Cauchy principal value scenario, this is genuine convergence) our conclusion is that

$$\int_{-\infty}^{\infty} f(x) \, dx = \frac{-2\pi i e^{\pi i a}}{1 - e^{2\pi i a}}$$

and it needs a little further work to confirm that this is actually a real number:

$$= \frac{-2\pi i}{\cos(\pi a) - i\sin(\pi a) - (\cos(\pi a) + i\sin(\pi a))} = \frac{\pi}{\sin(\pi a)}.$$

Conclusion
Provided that $0 < a < 1$,

$$\int_0^\infty \frac{x^{a-1}}{x+1}\,dx = \int_{-\infty}^\infty \frac{e^{at}}{e^t+1}\,dt = \frac{\pi}{\sin(\pi a)}.$$

VARIATION 4: NIBBLE THE BOX

If there is a *difficulty* of some kind on an edge of some rectangular contour, it may be good to skirt around that difficulty, much as we did in 10.4.6.

10.4.9 **Example**   Investigate, using Cauchy principal value if necessary, the integrals from zero to infinity and from minus infinity to infinity of the functions

$$g(x) = \frac{\cos(ax)}{e^x - e^{-x}} \quad \text{and} \quad h(x) = \frac{\sin(ax)}{e^x - e^{-x}} \qquad (a \in \mathbb{R}, 0 < a < 1).$$

Roughwork   If we combine the two given functions into

$$\frac{\cos(ax) + i\sin(ax)}{e^x - e^{-x}} = \frac{e^{iax}}{e^x - e^{-x}}$$

then there is reason to think that we ought to consider the complex function

$$f(z) = \frac{e^{iaz}}{e^z - e^{-z}}$$

which, because it resembles[8] the example in 10.4.8, suggests we try a long rectangular contour whose left and right edges we can allow to 'go to infinity' while we look for a limiting value. This problem is, however, significantly different from 10.4.8, because the function is undefined at the point 0 on the real axis—indeed, this is one of the reasons why the posed integrals are improper (and it will turn out to be the reason why we have to settle for Cauchy principal values this time). So let us impose tiny semicircular indentations onto such a rectangle to avoid hitting any singularity, and see what happens. Since the denominator of $f(z)$ is $e^z - e^{-z} = e^{-z}(e^{2z} - 1)$ which outputs the value zero whenever $1 = e^{2z} = e^{2x+i2y} = e^{2x}(\cos(2y) + i\sin(2y))$, that is, when $x = 0$ but $y = 0, \pm\pi, \pm2\pi, \pm3\pi, \ldots$, it may be prudent to skirt around more than one singularity.

---

[8] Especially if we rewrite the formula for $f(z)$ in the form

$$\frac{e^{(1+ia)z}}{e^{2z} - 1}$$

it looks as if the only significant difference from 10.4.8 is the change from 1 to -1 on the bottom line.

Solution   With f defined by

$$f(z) = \frac{e^{iaz}}{e^z - e^{-z}}$$

consider the simple closed contour Γ depicted in the diagram

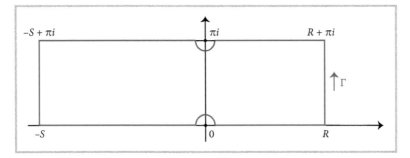

in which we take ε, the radius of the semicircular indentations by which we are side-stepping the singularities at 0 and at πi, to be positive and less than 1, but the numbers R and S to be both greater than 1. (This will at least prevent the semicircles from impacting the vertical edges.)

The function f has no singularities on or within Γ so, by the residue theorem (or, indeed, by Cauchy's theorem itself this time) we know that $\int_\Gamma f = 0$.

Let $I(R, S, \varepsilon)$ be the sum of the integrals along the intervals $[-S, -\varepsilon]$ and $[\varepsilon, R]$ on the real axis. We intend $I(R, S, \varepsilon)$ to give us an approximation for $\int_{-\infty}^{\infty} f(x)\, dx$ or its CPV, hoping to find a limit for this approximation as R and S tend to infinity and as ε tends to zero.

For each real x (except 0) notice that $f(x + \pi i) = -e^{-\pi a} f(x)$. It follows, just as in 10.4.8, that the integral of f along the (right-to-left) top side of Γ *excluding the semicircular detour* equals $e^{-\pi a} I(R, S, \varepsilon)$.

On the right-hand edge joining R to $R + \pi i$, described by $\gamma(t) = R + ti,\quad 0 \le t \le \pi$, the integral of f is

$$I(right, R) = \int_0^\pi \frac{e^{iaR - at}}{e^{R+it} - e^{-R-it}} i\, dt,$$

whose modulus is at most

$$\int_0^\pi \frac{e^0}{e^R - e^{-R}}\, dt$$

which certainly tends to zero as $R \to \infty$. So $I(right, R)$ must tend to 0 as $R \to \infty$ and, for like reasons, also the integral $I(left, S)$ along the left-hand edge must tend to zero as $S \to \infty$.

Now $e^z - e^{-z}$ has a Taylor expansion beginning

$$\left(1 + z + \frac{z^2}{2!} + \cdots \right) - \left(1 - z + \frac{z^2}{2!} - \cdots \right) = 2z + 2\frac{z^3}{3!} + \cdots$$

and so it has a *simple* zero at 0, giving f a *simple* pole there. This is what allows us to use 10.4.5 to access the integrals along the semicircular indentations: each has a limit of πi times the appropriate residue if anticlockwise, and therefore $-\pi i$ times the residue if clockwise (which, in fact, both of them are).

The residue of $f(z)$ at 0 is $\dfrac{1}{2}$ and the residue at πi is $-\dfrac{1}{2}e^{-\pi a}$.

Summarising what we now have yields the equation

$$0 = (1 + e^{-\pi a})I(R, S, \varepsilon) + I(right, R) + I(left, S) + (-\pi i)\left(\frac{1}{2} - \frac{1}{2}e^{-\pi a}\right).$$

Allowing R and S to tend to infinity independently,[9] and denoting by $I(\varepsilon)$ the resulting limit of $I(R, S, \varepsilon)$, we obtain

$$(1 + e^{-\pi a})I(\varepsilon) = \frac{\pi i}{2}(1 - e^{-\pi a})$$

and, since $I(\varepsilon)$ comprises the two (convergent improper) integrals

$$\int_{-\infty}^{-\varepsilon} \frac{\cos(ax) + i \sin(ax)}{e^x - e^{-x}}\, dx \text{ and } \int_{\varepsilon}^{\infty} \frac{\cos(ax) + i \sin(ax)}{e^x - e^{-x}}\, dx,$$

we can separate out the real and imaginary parts and let $\varepsilon \to 0$ *noting that this action is a CPV process:*

$$\text{CPV} \int_{-\infty}^{\infty} \frac{\cos(ax)}{e^x - e^{-x}}\, dx = 0, \qquad \text{CPV} \int_{-\infty}^{\infty} \frac{\sin(ax)}{e^x - e^{-x}}\, dx = \frac{\pi}{2}\left(\frac{1 - e^{-\pi a}}{1 + e^{-\pi a}}\right).$$

In retrospect, the first of these is obvious since the integrand is an odd function, and it tells us nothing about the integral from zero to infinity. The second is more informative, because the integrand is an even function and, indeed, has a limit (namely, $a/2$) as x approaches zero (via l'Hôpital); so in fact we have proper convergence here, and

---

[9] Therefore this is not CPV *at this stage*, but genuine convergence.

$$\int_{-\infty}^{0} \frac{\sin(ax)}{e^x - e^{-x}}\, dx = \int_{0}^{\infty} \frac{\sin(ax)}{e^x - e^{-x}}\, dx = \frac{\pi}{4}\left(\frac{1 - e^{-\pi a}}{1 + e^{-\pi a}}\right).$$

Postscript   When x is close to zero, $\cos(ax)$ is approximately 1, and $e^x - e^{-x}$ (as we saw above) is approximately $2x$. Also

$$\int_{\varepsilon}^{1} \frac{1}{2x}\, dx$$

diverges to infinity as positive $\varepsilon \to 0$. With a little more care this will allow us to show that

$$\int_{0}^{\infty} \frac{\cos(ax)}{e^x - e^{-x}}\, dx$$

is divergent, and the first of our CPV conclusions above is *necessarily only* a Cauchy principal value.

VARIATION 5: SAY 'CHEESE'

A wedge shape formed by two long straight lines sixty degrees apart coming out from the origin and closed off by an arc of a circle centred on 0 is an improbably cheesy shape around which to integrate, but sometimes it pays dividends.

10.4.10 **Example**   By investigating the integral of the function

$$f(z) = \frac{z}{1 + z^6}$$

around the illustrated (anticlockwise) contour Γ, evaluate the improper integral

$$\int_{0}^{\infty} \frac{x}{1 + x^6}\, dx.$$

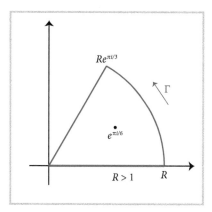

Solution Since f is regular except at the sixth roots of -1, only one of which (namely $e^{\pi i/6}$) lies inside Γ, we see that

$$2\pi i \operatorname{Res}(f, e^{\pi i/6}) = \int_{[0,R]} f + \int_{arc} f + \int_{slant} f$$

where the notation refers, respectively, to the horizontal portion of Γ, the curved part, and the slanting side. Now $\int_{arc} f$ tends to zero as $R \to \infty$ for virtually the same reasons as before, and $\int_{[0,R]} f$ is the real integral

$$I(R) = \int_0^R f(x)\, dx$$

whose limit is of interest to us. Next, to determine $\int_{slant} f$, notice that the path given by $\gamma(t) = te^{\pi i/3}, \quad 0 \le t \le R$ gives us the slanting edge *from bottom left to top right,* that is, the wrong way round. Therefore

$$\int_{slant} f = -\int_\gamma f(z)\, dz = -\int_0^R \frac{te^{\pi i/3}}{1 + (te^{\pi i/3})^6} e^{\pi i/3}\, dt = -e^{2\pi i/3} I(R).$$

Rearranging and letting $R \to \infty$, we now have $(1 - e^{2\pi i/3}) \lim I(R) = 2\pi i \operatorname{Res}(f, e^{\pi i/6})$

$$= 2\pi i \frac{z}{6z^5}\Big|_{z=e^{\pi i/6}} = \frac{\pi i}{3} \frac{1}{e^{2\pi i/3}} \quad \text{from which we get}$$

$$\int_0^\infty \frac{x}{1 + x^6}\, dx = \lim I(R) = \frac{\pi i}{3e^{2\pi i/3}(1 - e^{2\pi i/3})}$$

which routinely simplifies to $\dfrac{\pi\sqrt{3}}{9}$ (compare an earlier evaluation of this integral in paragraph 10.2.4, where the complex methods we were using there did not succeed in determining it).

10.5 The Gaussian integral–complex analysis showing off

'Everybody knows' that the function whose graph is the bell-shaped curve central to the whole of statistics

$$e^{-x^2}$$

cannot be integrated by elementary technology (that is, there is no *formula* in the usual sense of that word whose derivative is e^{-x^2}) and yet it is of the utmost importance that the improper integral

$$\int_{-\infty}^{\infty} e^{-x^2}\, dx$$

is convergent to a known value. We shall round off the chapter by handling this integral via the calculus of residues; the interested reader will find much fuller discussion on related matters in D Desbrow, *Amer Math Monthly* 105(8) (October 1998) 726–31 and in R Remmert's classic textbook *Theory of Complex Functions*.

The initial problem we face here is that the obvious starting-point complex function e^{-z^2} is regular everywhere and has no residues to feed into the machinery we have on hand. It is, however, possible to devise a function that does have residues and whose integrals along certain straight lines will give us what we need. The following choices are not at all obvious, but bear with us—all will be revealed.

Definition   Take a to be that square root of πi that has positive imaginary part. So $a^2 = \pi i$ and, to be precise, $a = (1 + i)\sqrt{\dfrac{\pi}{2}}$.

Definition   Take g to be the complex function described by the formula

$$g(z) = \frac{e^{-z^2}}{1 + e^{-2az}}.$$

We note that g has simple poles at the points where $e^{-2az} = -1$, an equation that solves easily enough to give $z = (n - \tfrac{1}{2})a$ for every integer n. It may be helpful to visualise where in the complex plane these poles lie:

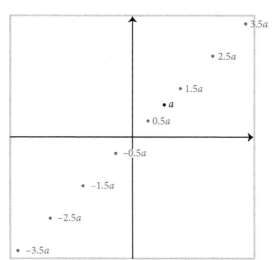

Here is the first indication that our seemingly random choices have an underlying battle plan.

10.5.1 Lemma   For every z except, of course, the poles of g, $g(z) - g(z+a) = e^{-z^2}$.

Proof   It is a routine calculation.   ∎

Definition   Let Q be the parallelogram-shaped contour (with anticlockwise orientation) having as vertices $-r, s, s + a, -r + a$ (and then back to $-r$) where r and s are large positive real numbers. (We propose to let both of these tend to infinity.) It is becoming more important to maintain a visual understanding of where this contour sits against the background of the previously identified poles of g:

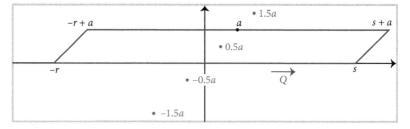

We note that the only pole of g that lies inside Q is $a/2$ and (keeping in mind that it is a *simple* pole) the usual f/g' formula (9.3.4) calculates its residue there; it simplifies to $-i/(2\sqrt{\pi})$. Next, we estimate the integral of g along the sloping side of Q from s to $s + a$: a line of length $|a|$.

While z lies on the segment $[s, s + a]$ we have $\text{Re}(z) \geq s$ and $\text{Im}(z) \leq \text{Im}(a)$, and it follows that

$$\left| e^{-z^2} \right| = e^{\text{Re}(-z^2)} = e^{-\,\text{Re}(z^2)}$$

$$= e^{-(\mathrm{Re}(z)^2 - \mathrm{Im}(z)^2)} \le e^{-s^2 + \mathrm{Im}(a)^2}.$$

Furthermore, for sufficiently large values of s we shall get $\left| e^{-2az} \right| \le \frac{1}{2}$, and consequently

$$\left| 1 + e^{-2az} \right| \ge 1 - \left| e^{-2az} \right| \ge \frac{1}{2}.$$

Hence the integral of g along $[s, s+a]$ cannot exceed $2|a|e^{-s^2 + \mathrm{Im}(a)^2}$, which tends to zero as $s \to \infty$.

For much the same reasons, the integral of g along the segment $[-r+a, -r]$ tends to zero as $r \to \infty$.

The integral along the base line of Q is, of course, $\int_{-r}^{s} g(x)\, dx$. The remaining item to check out carefully is the integral along the top line (which is, we notice, traced from right to left). Describing this as, for example,

$$\gamma(t) = -t + a, \quad -s \le t \le r,$$

with $\gamma'(t) = -1$, the top-line integral is

$$\int_{-s}^{r} g(-t+a)(-1)\, dt = \int_{s}^{-r} g(x+a)(-1)(-1)\, dx = \int_{s}^{-r} g(x+a)\, dx = -\int_{-r}^{s} g(x+a)\, dx$$

using the substitution $x = -t$. Now we are ready to invoke the residue theorem: 'two pi i times the (only) residue inside Q' is $\sqrt{\pi}$, and this has to equal

$$\int_{-r}^{s} g(x)\, dx - \int_{-r}^{s} g(x+a)\, dx + R_s + L_r = \int_{-r}^{s} \big(g(x) - g(x+a) \big)\, dx + R_s + L_r$$

where R_s and L_r are the integrals along the sloping sides of Q: quantities that we know to tend to zero as r and s tend to infinity independently (so this will not be a CPV, but proper convergence).

Lastly, using the formula in 10.5.1 to revamp the last display, we obtain

$$\sqrt{\pi} = \int_{-r}^{s} e^{-x^2}\, dx + R_s + L_r.$$

Letting r and s tend to infinity, we conclude that

$$\int_{-\infty}^{\infty} e^{-x^2}\, dx = \sqrt{\pi}.$$

10.6 Exercises

1. Suppose that $f(x) = \dfrac{p(x)}{q(x)}$ is any bottom-heavy rational (real) function whose denominator $q(x)$ has no real zeros. Prove that all of the improper integrals

$$\int_0^\infty f(x)\,dx, \quad \int_{-\infty}^0 f(x)\,dx, \quad \int_{-\infty}^\infty f(x)\,dx$$

are convergent. (*Hint:* begin by looking at the proof of 10.3.1.)
Why does it follow that

$$\int_0^\infty (a\cos(kx) + b\sin(kx))f(x)\,dx, \quad \int_{-\infty}^0 (a\cos(kx) + b\sin(kx))f(x)\,dx,$$

$$\int_{-\infty}^\infty (a\cos(kx) + b\sin(kx))f(x)\,dx$$

are also convergent for any real constants a, b and k?

2. Use contour integration to evaluate

$$I = \int_0^{2\pi} \frac{1}{17 + 8\cos\theta}\,d\theta.$$

3. Evaluate

$$\int_0^{\pi/2} \frac{1}{13 - 5\sin\theta}\,d\theta.$$

4. Use contour integration to evaluate

$$I = \int_0^{2\pi} \frac{1}{7 + 3\cos\theta - 2\sin\theta}\,d\theta.$$

5. Evaluate the improper integral

$$\int_{-\infty}^\infty \frac{x^2 - x + 2}{x^4 + 10x^2 + 9}\,dx.$$

6. Determine $\int_0^\infty \dfrac{x^2 + 5}{x^4 + 81}\,dx.$

7. Find the Cauchy principal value of

$$\int_{-\infty}^\infty \frac{\cos x}{x^4 + 4}\,dx.$$

8. Find the (Cauchy principal) value of

$$\int_0^\infty \frac{\cos x}{(x^2 + 1)(x^2 + 2)(x^2 + 3)} \, dx.$$

9. Find the (Cauchy principal) value of

$$\int_{-\infty}^\infty \frac{\cos(nx)}{(x^2 + a^2)^2} \, dx$$

where $n \in \mathbb{N}$ and $a > 0$.

10. Find the Cauchy principal value of

$$\int_{-\infty}^\infty \frac{x^5 \sin x}{x^6 + 64} \, dx.$$

11. Find the Cauchy principal values of

$$\int_{-\infty}^\infty \frac{x \cos(3x)}{x^2 + 4x + 5} \, dx \text{ and of } \int_{-\infty}^\infty \frac{x \sin(3x)}{x^2 + 4x + 5} \, dx.$$

12. Find the Cauchy principal value of

$$\int_0^\infty \frac{x \sin x}{1 - x^4} \, dx.$$

13. Evaluate

$$\int_0^\infty \frac{x^3 \sin x}{x^4 + 8x^2 + 16} \, dx.$$

14. Find the Cauchy principal value of

$$\int_{-\infty}^\infty \frac{x^2}{2401 - x^4} \, dx.$$

15. Evaluate the Cauchy principal value of

$$\int_{-\infty}^\infty \frac{x}{(x^3 - 8)(x^2 + 9)} \, dx.$$

16. Find the Cauchy principal value of

$$\int_0^\infty \frac{\cos x}{\cosh x} \, dx$$

by integrating the function $f(z) = \dfrac{e^{iz}}{\cosh z}$ around the rectangle with vertices at

$$R, R + i\pi, -R + i\pi, -R$$

(and back again to R, of course) where R is a large positive real number. You should find that the integral along each of the 'vertical' sides of the rectangle tends to zero as $R \to \infty$, and that the integral along the 'top' side is a multiple of the integral along the 'bottom' side.

17. Consider the same rectangular contour as described in the preceding exercise, but indented by a small semicircular detour around πi; also let

$$f(z) = \frac{ze^z}{e^{2z} - 1}.$$

Notice that, at 0, f has (only) a removable singularity, which we can 'remove' and effectively ignore.

By integrating f around the contour indicated, and verifying that the integrals along the vertical sides tend to zero whereas (in CPV terms) the integral along the top side has the same real part as the integral along the bottom, show that

$$\text{CPV} \int_{-\infty}^\infty \frac{xe^x}{e^{2x} - 1} \, dx = \frac{\pi^2}{4}.$$

18. Consider a contour as described in 10.4.10 but with an angle of $\pi/6$ (instead of $\pi/3$) at 0. For the function $f(z) = e^{-z^2}$, use integration of f around this contour to show that (in CPV terms)

$$\cos\left(\frac{\pi}{6}\right) \int_0^\infty e^{-x^2} \, dx = \int_0^\infty e^{-t^2/2} \cos\left(\sqrt{3}t^2/2\right) dt.$$

Hence evaluate the CPV of the latter integral.

11 The repair shop for broken promises

11.1 The field axioms

11.1.1 Definition   A set $\mathbb{F}$ of objects (which can be of any kind whatsoever, although it is convenient to think of them as 'numbers' in some sense) is called a *field* if there are two operations, usually called *addition* and *multiplication* defined upon $\mathbb{F}$ (and usually written in the familiar conventional notations) that satisfy all of the following conditions or *field axioms*:

1. **Addition axioms**
   (a) When a and b belong to $\mathbb{F}$ then $a + b$ belongs to $\mathbb{F}$
   (b) $a + b = b + a$ (for all a, b in $\mathbb{F}$)
   (c) $a + (b + c) = (a + b) + c$ (for all a, b, c in $\mathbb{F}$)
   (d) There is a special element '0' in $\mathbb{F}$ such that $a + 0 = a$ (for all a in $\mathbb{F}$)
   (e) For each $a \in \mathbb{F}$ there is a special element '$-a$' in $\mathbb{F}$ such that $a + (-a) = 0$

2. **Multiplication axioms**
   (a) When a and b belong to $\mathbb{F}$ then ab belongs to $\mathbb{F}$
   (b) $ab = ba$ (for all a, b in $\mathbb{F}$)
   (c) $a(bc) = (ab)c$ (for all a, b, c in $\mathbb{F}$)
   (d) There is a special element '1' in $\mathbb{F}$ such that $a1 = a$ (for all a in $\mathbb{F}$)
   (e) For each $a \in \mathbb{F} \setminus \{0\}$ there is a special element 'a^{-1}' in $\mathbb{F}$ such that $aa^{-1} = 1$

3. **The distributivity connection between addition and multiplication**
   (a) $a(b + c) = ab + ac$ (for all a, b, c in $\mathbb{F}$)

Common experience is that the real numbers $\mathbb{R}$ and the rational numbers $\mathbb{Q}$ do behave like this (that is, do form fields) and that these are the basic rules upon which we depend whenever we carry out routine, 'unthinking' arithmetic or algebra. Therefore whenever, in Chapter 2, we spoke of the complex numbers $\mathbb{C}$ also obeying *the usual arithmetical rules*, these were explicitly the rules or axioms that we had in mind.

Integration with Complex Numbers. McCluskey and McMaster, Oxford University Press.
© Brian McMaster and Aisling McCluskey (2022). DOI: 10.1093/oso/9780192846075.003.0011

11.2 L'Hôpital's rule for complex functions

11.2.1 Proposition   Suppose that f and g are regular on an open disc $D = D(a, r)$, that both functions have zeros at a and that, in particular, the zero of g at a is a simple zero. Then

$$\lim_{z \to a} \frac{f(z)}{g(z)} = \lim_{z \to a} \frac{f'(z)}{g'(z)}.$$

Proof   If f happens to be identically zero on D then so is f' and there is virtually nothing to prove. So we can assume that f has a zero of finite order, say, of order $m \geq 1$ at a. Taylor, and its associated 'factorisation at a zero', allow us to write

$$f(z) = (z - a)^m h(z) = (z - a)^m (a_0 + a_1(z - a) + a_2(z - a)^2 + \cdots)$$

where h is regular and non-zero on a (possibly smaller) disc. Equally,

$$g(z) = (z - a)^1 k(z) = (z - a)(b_0 + b_1(z - a) + b_2(z - a)^2 + \cdots)$$

where k is also regular and non-zero. Therefore (provided that $z \neq a$):

$$\frac{f(z)}{g(z)} = \frac{(z - a)^m h(z)}{(z - a) k(z)} = (z - a)^{m-1} j(z),$$

with $j(z) = h(z)/k(z)$ also being a non-zero regular function. From this we see that

- if $m > 1$ then $\dfrac{f(z)}{g(z)} \to 0$ (as $z \to a$) and also (via term-by-term differentiation)

$$\frac{f'(z)}{g'(z)} = \frac{m a_0 (z - a)^{m-1} + (m + 1) a_1 (z - a)^m + \cdots}{b_0 + 2 b_1 (z - a) + \cdots} \to \frac{0}{b_0} = 0$$

  since b_0 cannot be zero.

- if $m = 1$ then $\lim \dfrac{f(z)}{g(z)} = \lim \dfrac{h(z)}{k(z)} = \dfrac{h(a)}{k(a)} = \dfrac{a_0}{b_0} = \dfrac{f'(a)}{g'(a)}$, bearing in mind that all relevant functions are regular, non-zero and sums of power series.

   In both cases the limits set out in the statement of the proposition exist[1] and are equal.

11.2.2 Suggestion   Extend this result to show that, very much in line with 'real' calculus, consideration of $f'/g', f''/g'', f^{(3)}/g^{(3)}$ and so on until we break out of

---

[1] Contrast this with the 'real' version of l'Hôpital, in which the existence of a limit for f'/g' has to be declared as part of the hypotheses.

the zero-over-zero pathology will eventually yield the limit of f/g, provided that the order of zero that g has is finite and the order of zero that f has is at least as big. That is: given that f and g are regular on an open disc $D = D(a, r)$, that g has a zero of (finite) order m at a, and that f has a zero of order q at a where $q \geq m$, show that

$$\lim_{z \to a} \frac{f(z)}{g(z)} = \lim_{z \to a} \frac{f^{(m)}(z)}{g^{(m)}(z)} = \frac{f^{(m)}(a)}{g^{(m)}(a)}.$$

11.3 Swopping summation and integration

There is more than one meaningful way in which a sequence of *functions* can tend to a *function* as limit. We need to engage with two of them if we want to understand fully why and how Taylor's and Laurent's theorems work. The following definition applies equally to real functions and to complex functions but, of course, the main applications we have in mind are to the behaviour of complex functions.

11.3.1 Definition   Suppose that $(f_n)_{n \in \mathbb{N}}$ is a sequence of functions and that g is a function, and that T is a set lying within the domain of all of them. We say that

1. $(f_n)_{n \in \mathbb{N}}$ *converges pointwise on* T to g if, for each individual element t of T, we have $f_n(t) \to g(t)$);
2. $(f_n)_{n \in \mathbb{N}}$ *converges uniformly on* T to g if, for each $\varepsilon > 0$, there exists $n_\varepsilon \in \mathbb{N}$ such that $|f_n(t) - g(t)| < \varepsilon$ for every $t \in T$ and for every $n \geq n_\varepsilon$.

In the case where all relevant functions have the same domain T, we more simply say f_n converges pointwise to g in the first case, and f_n converges uniformly to g in the second.

It follows directly from the definitions that uniform convergence is *stronger* than pointwise convergence, that is, if f_n converges uniformly to g (on a set) then also f_n converges pointwise to g. The distinction between these two ideas is generally reckoned as quite tricky to grasp at first sight, so we shall flag up a few simple examples on real functions to try to render it more easily visible.

11.3.2 Example   For each positive integer n let us consider the real function $f_n : [0, 3] \to \mathbb{R}$ described by

$$f_n(x) = \begin{cases} nx & \text{if } 0 \leq x \leq \frac{1}{n}, \\ 2 - nx & \text{if } \frac{1}{n} \leq x \leq \frac{2}{n}, \\ 0 & \text{if } \frac{2}{n} \leq x \leq 3. \end{cases}$$

Also let g be the zero function on the same interval: $g(x) = 0$ for each $x \in [0, 3]$.

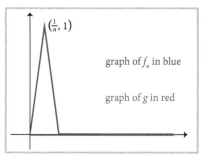

$(\frac{1}{n}, 1)$

graph of f_n in blue

graph of g in red

It is routine to verify that f_n tends to g pointwise, because

1. if $x = 0$, then every $f_n(x)$ is zero, whereas
2. if $0 < x \leq 3$ then the 'spike' in the graph of f_n eventually moves completely to the left of x as n increases, guaranteeing that $f_n(x)$ shall equal zero for all sufficiently large values of n.

On the other hand, the presence of a spike of height 1 on the graph of every f_n tells us that, no matter which n_1 we think of, there will be values of n greater than that such that the condition $|f_n(x) - g(x)| < 1$ will fail at the 'summit of the spike' for infinitely many $n \geq n_1$; so f_n does not converge uniformly to g.

Essentially the same style of argument will demonstrate each of the following:

11.3.3 Example   For each positive integer n let us consider the real function $f_n : [0, 3] \to \mathbb{R}$ described by

$$f_n(x) = \begin{cases} n^2x & \text{if } 0 \leq x \leq \frac{1}{n}, \\ 2 - n^2x & \text{if } \frac{1}{n} \leq x \leq \frac{2}{n}, \\ 0 & \text{if } \frac{2}{n} \leq x \leq 3. \end{cases}$$

Also let g be the zero function on the same interval: $g(x) = 0$ for each $x \in [0, 3]$. Then f_n converges pointwise to g but f_n does not converge uniformly to g. The same graph as in 11.3.2 will serve to illustrate this example also: the only change is that the height of the spike now increases with n.

11.3.4 Example   For each positive integer n let us consider the real function $f_n : [0, 3] \to \mathbb{R}$ described by

$$f_n(x) = \begin{cases} 1 - nx & \text{if } 0 \leq x \leq \frac{1}{n}, \\ 0 & \text{if } \frac{1}{n} \leq x \leq 3. \end{cases}$$

Also let g be function described by

$$g(x) = \begin{cases} 1 & \text{if } x = 0, \\ 0 & \text{if } 0 < x \le 3. \end{cases}$$

Then f_n converges pointwise to g but f_n does not converge uniformly to g.

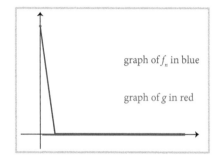

graph of f_n in blue

graph of g in red

11.3.5 Note

- In Example 11.3.4, even though every f_n was continuous, their pointwise limit g was not.
- In Example 11.3.3 it is easy to see that (for every n) $\int_0^3 f_n(x)\, dx = 1$ but that $\int_0^3 g(x)\, dx = 0$: so although f_n converges pointwise to g, the integral of f_n does not converge to the integral of g.

These two facts should serve as a warning: if we intend to do any serious analysis on a sequence of functions, involving perhaps continuity and integration, then pointwise convergence is unlikely to be a suitable tool with which to do it. In contrast, it turns out that *uniform* convergence does respect both continuity and integrals.

11.3.6 Proposition

1. If the sequence $(f_n)_{n \in \mathbb{N}}$ of real functions converges uniformly to (continuous) g on $[a, b]$, then

$$\int_a^b f_n(x)\, dx \to \int_a^b g(x)\, dx.$$

2. If the sequence $(f_n)_{n \in \mathbb{N}}$ of complex functions converges uniformly to (continuous) g on the track of a contour γ, then

$$\int_\gamma f_n(z)\, dz \to \int_\gamma g(z)\, dz.$$

Proof In the case of part (1), given $\varepsilon > 0$, uniform convergence tells us that there is $n_0 \in \mathbb{N}$ such that, for $n \geq n_0$, $|f_n(x) - g(x)| < \dfrac{\varepsilon}{b-a}$ at every point of $[a, b]$. Therefore

$$\left| \int_a^b f_n(x)\, dx - \int_a^b g(x)\, dx \right| \leq \int_a^b |f_n(x) - g(x)|\, dx \leq \int_a^b \left(\frac{\varepsilon}{b-a} \right) dx = \varepsilon.$$

The proof of part (2) is essentially the same but uses the length of γ in place of the length $b - a$ of the interval $[a, b]$. ∎

11.3.7 Exercise Verify that if a sequence of (real or complex) continuous functions converges uniformly to a limit function, then that limit function is also continuous.

The ideas of pointwise and uniform convergence transfer over to a *series of functions* in just the same way that we see in elementary analysis for series of numbers—one simply focuses on its sequence of partial sums:

11.3.8 Definition Suppose that $\sum_{k=1}^{\infty} w_k$ is a series of (real or complex) functions and that g is a function, and that T is a set lying within the domain of all of them. We say that

1. $\sum w_k$ *converges pointwise on T to g* if the sequence $(S_n)_{n \in \mathbb{N}}$ of partial sums, defined by

$$S_n = w_1 + w_2 + w_3 + \cdots + w_n = \sum_1^n w_k,$$

converges pointwise on T to g;

2. $\sum w_k$ *converges uniformly on T to g* if the sequence $(S_n)_{n \in \mathbb{N}}$ of partial sums converges uniformly on T to g.

Of course it is legitimate to start off a series of functions at $k = 0$ in place of $k = 1$, and this is in fact the norm for power series in particular.

The earlier result (11.3.6) concerning integration of a uniform limit applies to series sums in the way we need in order to swop summation and integration:

11.3.9 Proposition Suppose that $g(z) = \sum_{k=1}^{\infty} w_k(z)$ where the convergence takes place on a domain D, and that the convergence is uniform on the track of a contour γ in D. Then

$$\int_\gamma \sum_{k=1}^\infty w_k(z)\, dz = \sum_{k=1}^\infty \int_\gamma w_k(z)\, dz;$$

that is, we can interchange the order of integration and summation provided that the convergence is uniform.

Proof   Putting $S_n(z) = \sum_1^n w_k(z)$ for each $n \in \mathbb{N}$ and each $z \in D$, what we are told is that $S_n(z) \to g(z)$ throughout D and that $S_n(z) \to g(z)$ uniformly on $T =$ the track of γ. By 11.3.6 we then have $\int_\gamma S_n(z)\, dz \to \int_\gamma g(z)\, dz$, that is,

$$\int_\gamma \sum_{k=1}^\infty w_k(z)\, dz = \lim_{n \to \infty} \int_\gamma \sum_1^n w_k(z)\, dz$$

$$= \lim_{n \to \infty} \sum_1^n \int_\gamma w_k(z)\, dz = \sum_1^\infty \int_\gamma w_k(z)\, dz,$$

where in the last line we have used the fact (see, for instance, 6.6.8's first bullet point, which extends easily by induction) that integral *does* distribute over *finite* sums.   ∎

Here is one of the most useful results for confirming uniform convergence, generally known as the *Weierstrass M-test*:

11.3.10 Proposition   Suppose that $\sum_1^\infty w_k$ is a series of (real or complex) functions defined (at least) on a set T, that we can find (for each $k \in \mathbb{N}$) a positive real constant M_k such that

$$|w_k(t)| \le M_k \ \text{ for every } t \in T,$$

and that $\sum_1^\infty M_k$ is a convergent series. Then $\sum_1^\infty w_k$ is uniformly convergent (to some function) on T.

Proof   For each $t \in T$, the direct comparison test tells us that $\sum |w_k(t)|$ is convergent, that is, $\sum w_k(t)$ is absolutely convergent and consequently convergent. Therefore the definition

$$g(t) = \sum_1^\infty w_k(t) \ \text{ for each } t \in T$$

makes sense, and defines a function g on the set T.

Now $\sum_1^\infty M_k$ is convergent to a sum-to-infinity—let us denote it by Q—and so for any given $\varepsilon > 0$ we can find $n_\varepsilon \in \mathbb{N}$ such that

$$\sum_{n+1}^\infty M_k = \left| Q - \sum_1^n M_k \right| < \varepsilon \quad \text{for every } n \geq n_\varepsilon.$$

For any $t \in T$ and any $n \geq n_\varepsilon$ we find that

$$\left| \sum_1^n w_k(t) - g(t) \right| = \left| \sum_{n+1}^\infty w_k(t) \right| \leq \sum_{n+1}^\infty |w_k(t)| \leq \sum_{n+1}^\infty M_k < \varepsilon$$

as required for uniform convergence of $\sum_1^\infty w_k$ to g.   ∎

A slick and important application of the Weierstrass M-test is to show that any power series will be *uniformly* convergent on a set that (in an appropriate sense) 'stays well inside' its circle of convergence. Here is a relevant particular case:

11.3.11 Proposition   Let $\sum_0^\infty a_k(z - a)^k$ be a complex power series that converges at every point in a disc $D(a, r)$, and let $C = C(a, r')$ be a circle centre a whose radius r' is strictly smaller than r. Then the power series is uniformly convergent on C.

Proof   Pick a point z_1 that is within $D(a, r)$ but is outside the circle C: that is, the distance $r'' = |z_1 - a|$ is greater than r' but less than r. Notice that the positive real number $\dfrac{r'}{r''}$ is less than 1.

Now $\sum a_k(z_1 - a)^k$ is convergent, so certainly $a_k(z_1 - a)^k \to 0$ and is therefore bounded, and we can find a constant B such that $|a_k(z_1 - a)^k| < B$ for every k. It follows that, for z on the circle C,

$$|a_k(z - a)^k| = \left| a_k \left(\frac{z - a}{z_1 - a} \right)^k (z_1 - a)^k \right| = |a_k(z_1 - a)^k| \left| \frac{r'}{r''} \right|^k \leq B \left| \frac{r'}{r''} \right|^k.$$

Since $\sum B \left| \dfrac{r'}{r''} \right|^k$ converges, the Weierstrass M-test gives uniform convergence on C of the original series.   ∎

Only slight modifications to this argument are necessary in order to get an analogous result for *double* power series of the 'Laurent' type:

11.3.12 Proposition   Let $\sum_{-\infty}^\infty a_k(z - a)^k$ be a (complex) double power series that converges at every point in a punctured disc $D(a, r) \setminus \{a\}$, and let $C = C(a, r')$

be a circle centre a whose radius r' is strictly smaller than r. Then the double power series is uniformly convergent on C.

11.3.13 Summary The essence of 11.3.9, 11.3.11 and 11.3.12 is that, when integrating either a Taylor series or a Laurent series around a circle that lies within its disc of convergence, it is legitimate to swop integral and summation. Now let's make use of that.

11.3.14 Theorem Laurent coefficients are unique in the following sense: if $\sum_{-\infty}^{\infty} a_k(z-a)^k$ and $\sum_{-\infty}^{\infty} b_k(z-a)^k$ converge to the same function on a punctured disc $D(a, r) \setminus \{a\}$, then $a_n = b_n$ for every integer (positive, negative or zero) n.

Proof Subtracting the two series we find that $f(z) = \sum_{-\infty}^{\infty} (a_k - b_k)(z - a)^k$ where f is the identically-zero function on the punctured disc. Pick any integer m and multiply $f(z)$ by $(z - a)^m$ and, of course, you still get zero everywhere, so $\int_C f(z)(z-a)^m \, dz = 0$ for any circle C centred on a and lying within that punctured disc. That is,

$$0 = \int_C f(z)(z-a)^m \, dz = \int_C \sum_{-\infty}^{\infty} (a_k - b_k)(z-a)^{m+k} \, dz = \sum_{-\infty}^{\infty} \int_C (a_k - b_k)(z-a)^{m+k} \, dz.$$

Yet the last named integral is easy: (see, for instance, 6.6.3) it is $2\pi i(a_k - b_k)$ if $m + k = -1$ (that is, when $k = -m - 1$) and it is zero in every other case. Therefore the last display collapses to

$$0 = 2\pi i(a_{-m-1} - b_{-m-1}).$$

Since m was free to be any integer, the result follows. ■

Now we can address the major *lacunae* in Chapters 8 and 9.

11.3.15 Filling the gap in Taylor (paragraph 8.2.3) The step in the proof of Taylor's theorem that we did not fully justify in Chapter 8 was

$$\int_C f(w) \sum_{0}^{\infty} \frac{(z-a)^n}{(w-a)^{n+1}} \, dw = \sum_{0}^{\infty} \int_C f(w) \frac{(z-a)^n}{(w-a)^{n+1}} \, dw$$

where f was continuous, C was a circle centred on a and, while the controlling variable w for the integration varied around C, the modulus of the ratio $\left(\dfrac{z-a}{w-a}\right)$ was actually a positive constant (call it t) *less than* 1.

Rewriting this slightly as

$$\int \sum f(w) \frac{(z-a)^n}{(w-a)^{n+1}} \, dw = \sum \int f(w) \frac{(z-a)^n}{(w-a)^{n+1}} \, dw,$$

all we need confirm before appealing to 11.3.9 to swop over the sum and the integral is that the series of functions converges uniformly on the circle C. Now the continuous function $|f(w)|$ possesses a biggest value (call it B) on C so, for each value of n,

$$\left| f(w) \frac{(z-a)^n}{(w-a)^{n+1}} \right| \leq B \left| \frac{z-a}{w-a} \right|^n \frac{1}{|w-a|} = \frac{B}{|w-a|} t^n.$$

Since the geometric series $\sum \frac{B}{|w-a|} t^n$ converges, the Weierstrass M-test guarantees the desired uniform convergence.

11.3.16 Filling the gap in Laurent (paragraph 9.1.1)   The gap in the proof of Laurent's theorem as we presented it in Chapter 9 was how, in detail, to reshape the integral that we called

$$-I_2 = - \int_{C^-} \frac{f(w)}{w-z} \, dw$$

into

$$\sum_{-\infty}^{-1} a_n (z-a)^n$$

with the coefficients specified as in 9.1.1, keeping in mind that C^- is a circle centred at a but of radius small enough that z lies outside it.

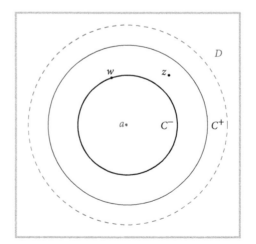

Now while w traverses C^-,

$$-(w - z) = (z - a) - (w - a) = (z - a)\left(1 - \frac{w - a}{z - a}\right) \quad \text{where} \quad \left|\frac{w - a}{z - a}\right| < 1,$$

and therefore

$$-\frac{1}{w - z} = \frac{1}{z - a}\left(1 - \frac{w - a}{z - a}\right)^{-1} = \frac{1}{z - a}\sum_{r=0}^{\infty}\left(\frac{w - a}{z - a}\right)^{r}$$

which gives

$$-I_2 = \int_{C^-} f(w) \sum_{r=0}^{\infty} \frac{(w - a)^r}{(z - a)^{r+1}}\, dw = \int_{C^-} \sum_{r=0}^{\infty} f(w) \frac{(w - a)^r}{(z - a)^{r+1}}\, dw.$$

At this point we want to swop over the summation and integration and, once again, all we need to verify is that the series of functions is uniformly convergent. Letting B denote the biggest value of $|f|$ on C^-, we have

$$\left| f(w) \frac{(w - a)^r}{(z - a)^{r+1}} \right| \le \frac{B}{|z - a|} \left| \frac{w - a}{z - a} \right|^r$$

and, since the geometric series $\sum \frac{B}{|z-a|}\left|\frac{w-a}{z-a}\right|^r$ converges, the Weierstrass M-test delivers the uniformity of convergence we sought.

Interchanging sum and integral, we now find

$$-I_2 = \sum_{r=0}^{\infty} \int_{C^-} f(w) \frac{(w - a)^r}{(z - a)^{r+1}}\, dw$$

(substituting $s = -r - 1$, that is, $r = -1 - s$)

$$= \sum_{s=-1}^{-\infty} \int_{C^-} f(w) \frac{(w - a)^{-1-s}}{(z - a)^{-s}}\, dw$$

$$= \sum_{s=-1}^{-\infty} \left(\int_{C^-} \frac{f(w)}{(w - a)^{s+1}}\, dw \right) (z - a)^s = \sum_{s=-1}^{-\infty} a_s (z - a)^s, \text{ say,}$$

which is the familiar 'canonical' form of the principal part of the Laurent expansion.

11.4 Smoothness: analytical and geometrical

One of the recurring themes of complex analysis is the interplay between analysis and geometry, and how they frequently support one another by bringing complementary insights to the same situation. It needs to be borne in mind, however, that the information content of these insights may not always be identical: in particular, diagrams are often useful more as a guide to what analysis to do, rather than as a complete alternative working strategy.

A case in point is how we determine whether a given path is smooth or not smooth. Let us pin the discussion down to something explicit by considering the following question, which is largely a recap from Chapter 6 Exercise 3:

11.4.1 Example   Decide whether or not the following paths are smooth:

1.

$$
\gamma_1(t) = \begin{cases} t - i & 0 \le t \le 2, \\ 2 + (t - 3)i & 2 \le t \le 4, \\ 6 - t + i & 4 \le y \le 6, \\ (7 - t)i & 6 \le t \le 8; \end{cases}
$$

2.

$$
\gamma_2(t) = \begin{cases} 1 + 2e^{-it} & -\pi \le t \le 0, \\ 3 - 2it & 0 \le t \le \frac{5}{2}; \end{cases}
$$

3.

$$
\gamma_3(t) = \begin{cases} 1 + 2e^{-\pi it} & -1 \le t \le 0, \\ 3 - it & 0 \le t \le 5. \end{cases}
$$

Solution

1. In the case of γ_1, a decent sketch diagram produces all the evidence we need for a firm conclusion. The sharp corners on the square-shaped track make it manifest that γ_1 cannot possibly be smooth, since each represents a point at which the direction of travel changes abruptly, and therefore the argument angles of the one-sided derivatives cannot be equal.

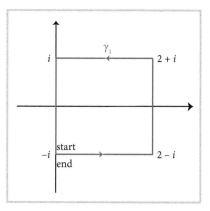

2. In the case of γ_2, the diagram shows no such sharpness of corner. As far as the geometric representation of the track can take us, this appears to be a smooth path; certainly it is one in which the direction of travel does not break discontinuously at the point where the two components of the track meet up, so the *argument angles* of the one-sided derivatives must agree.

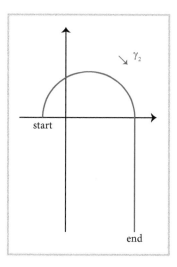

Yet the analytic definition of smoothness requires that we go further, and investigate also the *modulus* of the one-sided derivatives—the speeds of travel on each side of the junction. Noting that $\gamma_2'(t)$ is $-2ie^{-it}$ for $-\pi < t < 0$ but $-2i$ for $0 < t < \frac{5}{2}$, it is routine to check that $\gamma_2'(0) = -2i$ and that γ_2' is continuous throughout its domain. So γ_2 really is smooth.

3. Now consider γ_3. The track of γ_3, its starting and ending points, its direction of travel and the order in which it moves through the points on its track are identical to those of γ_2: 'geometry cannot see any difference between γ_2 and γ_3, and therefore believes that both are smooth.' Analysis, on the other

hand, wants investigation of speed as well as of direction, and here a critical difference emerges: for $\gamma_3'(t)$ is $-2\pi i e^{-\pi i t}$ for $-1 < t < 0$ but $-i$ for $0 < t < \frac{5}{2}$ and therefore, since

$$\lim_{t<0,t\to 0} \gamma_3'(t) = -2\pi i \neq -i = \lim_{t>0,t\to 0} \gamma_3'(t),$$

we see that it is impossible for γ_3' to be continuous on its domain even before we consider trying to assign a value to $\gamma_3'(0)$. (In effect, the speed of travel drops instantaneously from 2π units per second to one unit per second where the semicircular part of the track joins the straight final section.) We are forced to conclude that γ_3 is not smooth, even though geometry *doesn't notice*.

The reason why we have felt able to turn a blind eye to such an internal mis-alignment in two of the fundamental tools of the discipline up to the very last page of our book is simply this: *from the point of view of doing integrals, it doesn't matter*. The declared purpose of this text is to explore integration involving com-plex numbers, which essentially means integrating along paths. As Chapter 6 (see paragraph 6.6.18) stressed, such path integrals are unaffected by 'speed of trav-el' so, *firstly*, if we had reason to integrate a continuous function f along γ_3, we could (had we happened to notice its lack of smoothness) have altered its speed parameter on semicircular arc or straight line or both in order to fix the perceived problem without disturbing the eventual result and, *secondly*, this would have been a complete waste of time: for

$$\int_{\gamma_3} f = \int_{semicircle} f + \int_{straight} f$$

and neither item on the right-hand side is affected by changes of parameter anyway. In effect, what geometry tells us about smoothness is perfectly good enough for our primary purpose, as well as being more easily accessible in many cases.

Suggestions for further or supplementary reading

Ahlfors, L. *Complex Analysis: An Introduction to the Theory of Analytic Functions of One Complex Variable.* McGraw-Hill (1979).

Beck, M. et al. *A First Course in Complex Analysis*, 2nd edn. Orthogonal Publishing L3C (2018).

Churchill, R.V. and Brown, J.W. *Complex Variables and Applications.* McGraw-Hill (2013).

Dyer, R.H. and Edmunds, D.E. *From Real to Complex Analysis.* Springer Undergraduate Mathematics Series (2014).

Howie, J.M. *Complex Analysis.* Springer Undergraduate Mathematics Series (2008).

McCluskey, A. and McMaster, B. *Undergraduate Analysis—A Working Textbook.* Oxford University Press (2018).

Needham, T. *Visual Complex Analysis.* Oxford University Press (1997, reprinted 2012).

Norton, R. *Complex Variables for Scientists and Engineers—An Introduction.* Oxford University Press (2010).

Priestley, H.A. *Introduction to Complex Analysis*, 2nd edn. Oxford University Press (2003).

Rudin, R. *Real and Complex Analysis.* Walter Rudin Student Series in Advanced Mathematics (1985).

Stein, E.M. and Shakarchi, R. *Complex Analysis, Princeton Lectures in Analysis II.* Princeton Press (2007).

Stewart, I. and Tall, D. *Complex Analysis*, 2nd edn. Cambridge University Press (2018).

Yue Kuen Kwok. *Applied Complex Variables for Scientists and Engineers.* Cambridge University Press (2010).

Index